# Modeling of Z-pinned Carbon Fiber-Reinforced Polymer (CFRP) Composite

Yingjie Xu · Weihong Zhang

# Modeling of Z-pinned Carbon Fiber-Reinforced Polymer (CFRP) Composite

## Curing Process and Mechanical Properties

上海交通大學出版社
SHANGHAI JIAO TONG UNIVERSITY PRESS

Springer

Yingjie Xu
Northwestern Polytechnical University
Xi'an, Shaanxi, China

Weihong Zhang
Northwestern Polytechnical University
Xi'an, Shaanxi, China

ISBN 978-981-96-3627-3 ISBN 978-981-96-3628-0 (eBook)
https://doi.org/10.1007/978-981-96-3628-0

Jointly published with Shanghai Jiao Tong University Press
The print edition is not for sale in China (Mainland). Customers from China (Mainland) please order the print book from: Shanghai Jiao Tong University Press.



This Springer imprint is published by the registered company Springer Nature Singapore Pte Ltd.
The registered company address is: 152 Beach Road, #21-01/04 Gateway East, Singapore 189721, Singapore

# Preface

Z-pinning is a promising technique to improve interlaminar fracture toughness in composite laminates by providing through-thickness reinforcement. The Z-pinning technique offers several advantages, including its simple design, low cost, and applicable to prepreg materials. It has been used in Formula 1 racing car, Globemaster III heavy-lift transport aircraft and F/A-18 fighter aircraft. To design components satisfying complex service conditions, one should understand well the mechanical properties of Z-pinned composites that depend upon a variety of factors, e.g., processing conditions and Z-pining parameters, etc.

In recent years, our research group has been devoted to the development of experimental and numerical modeling techniques for mechanical properties of Z-pinned carbon fiber-reinforced polymer (CFRP) composites. The main content of this book contains five chapters covering experimental characterization, material modeling, FE simulation as well as the integrated analysis method for Z-pinned CFRP composites.

Chapter 1 briefly introduces the mechanical properties, processing, and applications of Z-pinned CFRP composites. The purpose and layout of this book are also outlined.

Chapter 2 presents the experiment methods used in characterizing the mechanical properties of Z-pinned CFRP composites.

Chapter 3 focuses on the modeling of the curing process of Z-pinned CFRP composites and evolution of cure-induced residual stresses. The effect of the cure-induced residual stresses on the interfacial cracking of Z-pins after curing is discussed in detail.

Chapter 4 presents a multiscale framework to evaluate the delamination behavior of Z-pinned CFRP composites including the effects of the cure-induced Z-pin/composite interfacial cracks.

Chapter 5 focuses on the development of an integrated computing framework that incorporates the curing defects into the mechanical simulation of Z-pinned CFRP composites.

This book is not indented to capture all the significant contributions in the literature. It can be considered as a complementary work of existing books about composite mechanics and processing. This book can be used as a guideline for graduate students and research engineers who wish to learn the basic experimental and modeling techniques in characterizing and designing Z-pinned composite products. Readers need the technical background of material science, solid mechanics, and finite element method.

Xi'an, China
June 2024

Yingjie Xu
Weihong Zhang

# Contents

# Chapter 1
# Introduction

## 1.1 Advantages of Z-pinned CFRP Composites

Carbon fiber reinforced polymer (CFRP) composites have been increasingly adopted as substitutes for metallic materials in a wide range of high-performance structural applications owing to their excellent mechanical and physical properties, such as high specific strength and stiffness, fatigue properties and strong design tolerability [1–3]. In most of the CFRP composites, laminate piles are stacked together with orientated fibers to provide in-plane reinforcement. A significant shortcoming of this process is that no fibers are arranged through the thickness direction. Therefore, CFRP composite laminates usually present low interlaminar resistance and show notorious susceptibility to delamination.

Z-pinning is a well-established through-thickness reinforcement (TTR) technology to improve the interlaminar properties of fiber reinforced polymer composite laminates. Z-pins are thin fiber or metallic rods with high stiffness and strength, as shown in Fig. 1.1 [1].

The Z-pinning technique presents many advantages, as shown in Fig. 1.2. Firstly, Z-pinning is the only technique capable of reinforcing laminated composites. Other reinforcements, such as stitching, 3D weaving, tufting, and braiding, are only applicable for textile laminates and require complete reinforcement through the thickness of the dry fabric preforms before resin infusion [2–6].

In addition, Z-pinning shows a higher toughening effect. Only a minimal content of Z-pins (typically 0.5–5.0 vol%) is required to significantly enhance the interlaminar properties of composites. Extensive studies have verified the reinforcement effects of Z-pining on fracture toughness, impact damage tolerance and interlaminar strength of composite laminates [7–11]. For example, the mode I delamination resistance of Z-pinned composites laminates (up to 10–15 kJ/m$^2$) is larger than that of composites reinforced with particles or short fibers (typically below 4 kJ/m$^2$) [12–16], stitches (5–10 kJ/m$^2$) [17] or 3D woven (up to 3–4 kJ/m$^2$) [18, 19], as shown in Table 1.1.



Y. Xu and W. Zhang, *Modeling of Z-pinned Carbon Fiber-Reinforced Polymer (CFRP) Composite*, https://doi.org/10.1007/978-981-96-3628-0_1

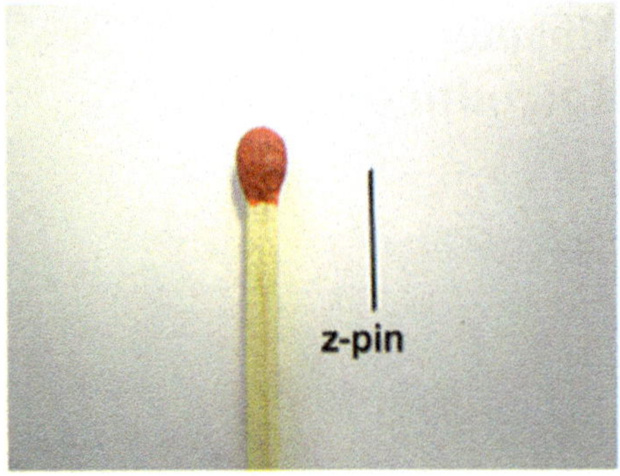


**Fig. 1.1** A typical Z-pin [1]

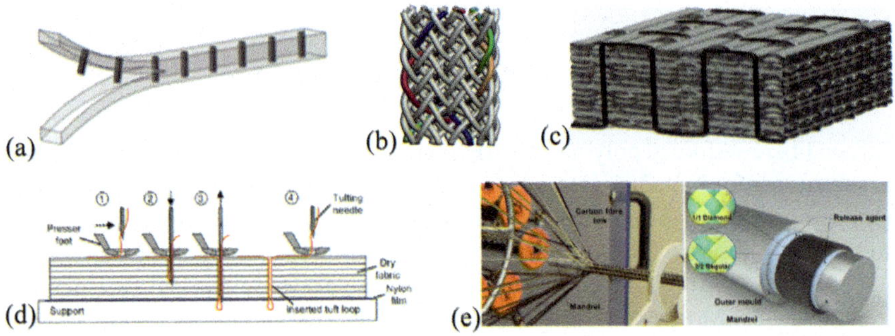


**Fig. 1.2** Schematic of different toughening method: **a** Z-pining, **b** stitching, **c** 3D weaving, **d** tufting, **e** braiding

Therefore, Z-pinning shows wider application prospects and research attractions than other TTR technologies.

To date, Z-pinning has been applied to strengthen the composite structural components on the Formula 1 racing cars, F/A-18 Super Hornet fighter aircraft, Globemaster III heavy-lift transport aircraft, Airbus A340-600, Boeing 7E7 and HOPE-X [20, 21]. The manufacturing process, structural characterization and functional properties of Z-pinned composites have been thoroughly reviewed [22].

**Table 1.1** Comparison of the improvement in mode I interlaminar property by different TTR techniques

| TTR techniques | Degree of improvement |
|---|---|
| Z-pin | ~ 1200% [3] |
| Stitching | ~ 750% [17] |
| 3D weaving | ~ 800% [4] |
| Tufting | ~ 1000% [5] |

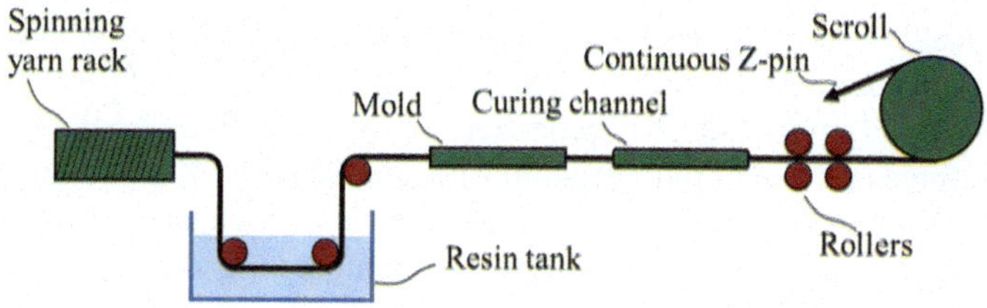


**Fig. 1.3** Schematic diagram of twisted continuous Z-pin pultrusion process

## 1.2 Processing of Z-pinned CFRP Composites

### *1.2.1 Type of Z-pin*

In addition to the traditional Z-pins made up of carbon fiber composite or metals, different types of Z-pins have been developed to increase the flexibility and availability of the design and manufacturing of Z-pinned composites. The major criteria for choosing Z-pins range from preferred reinforcement properties, matching with the host material, minimized in-plane property reduction, low cost and ease of application, etc. [23].

The most employed Z-pin is the cylindrical tow of carbon fibers/resin composites. Figure 1.3 presents the schematic diagram of twisted continuous Z-pin pultrusion process. Continuous carbon fiber is protruded through the liquid polymer at first. Sometimes, twisted fibers are chosen to increase the pullout force of Z-pins [24, 25]. The infiltrating fibers are then forced to pultrude and then cured at elevated temperatures. Bismaleimide (BMI) is the commonly-chosen resin since it can retain high mechanical properties under the elevated curing temperature [23].

Besides carbon fiber Z-pins, Z-pins made up of steel, titanium alloy and copper are also applied because of their high strength or electrical conductivity. Compared with the carbon fiber Z-pins, metal Z-pins present fewer defects, while the extra surface treatment for metal Z-pins is required to achieve acceptable adhesion to the composite host. Pure polymer Z-pins are rarely applied as their stiffness and strength are quite low [26].

### *1.2.2 Ultrasonically Assisted Z-fiber (UAZ) Process*

The most common method for the Z-pining is the UAZ (Ultrasonically assisted Z-Fiber) developed by Aztex Inc. (Waltham, USA) [27]. Figure 1.4 shows the schematic diagram of the UAZ process for manufacturing the Z-pinned composites. The compressible preform is firstly manufactured by inserting numerous Z-pins in the designed pattern. The compressible foam carrier ensures the even distribution of Z-pins and provides the necessary support during the Z-pin insertion process. The

Z-pinning pattern is determined by the spacing and content of Z-pins in this preform. This preform is then placed over the composite laminates. The high-frequency ultrasonic vibrations (around 20 kHz) generated by the actuated tool collapse the foam carrier and forces the Z-pins into the composite. The excessive Z-pins and compressed foam carriers are then shaved away from the composite laminates by blades. The Z-pinned composites are finally cured at elevated temperature and pressure to achieve high performance.

It should be noted that the Z-pins inserted through the UAZ process are often inclined up to 20–30° in the through-thickness direction, even though the Z-pins in the foam carrier are perfectly orthogonal [29–31]. The possible mechanisms of Z-pin misalignment come from the following issues:

- Firstly, the high-frequency ultrasonic vibrations from the tool inevitably compel the Z-pins to swing along the fibers.
- Secondly, the shearing action of the protruding Z-pins forces the Z-pins embedded in the composites to incline and rotate.
- Thirdly, the pressure during the curing process compacts the softened laminates at elevated temperatures and forces the rigid Z-pins to slide.
- Finally, the flowing process of resin also results in the misalignment of Z-pins.

For UAZ technique, Z-pins are hammered into the prepreg by ultrasonic horn. This process generally leads to severe fiber crimp, breakage, waviness and resin-rich zone. These initial defects introduced by Z-pin implantation would degrade the mechanical properties of Z-pinned composites. For instance, the in-plane tensile strength decreases linearly with the increase of volume content and diameter of Z-pins. The main reason for the decrease in tensile strength is fiber breakage. When pins are forced into the laminate, fiber breakage occurs around each pin. The broken fibers are the locations for the crack generation and tensile failures. This leads to a reduction in tensile strength.

### *1.2.3 Pre-hole Z-pinning (PHZ) Process*

The UAZ technique was the firstly developed and the most popular method for the manufacturing of Z-pinned CFRP composites. However, the existing work shows that UAZ process leads to a reduction of the in-plane properties influencing the ultimate strength of laminates. Therefore, pre-hole Z-pinning (PHZ) process has been developed to modify the insertion process based on the UAZ technique.

For the PHZ technique, Z-pins are produced by pulling continuous polyimide fiber filaments through resin and then pultruding through a circular mold—a transparent plastic tube with the same cross section as the Z-pin. As detailed in Fig. 1.5, the key steps of PHZ process are as follows:

(1) heat the prepreg stack at a temperature of 30 ± 3 °C to soften the resin matrix;
(2) penetrate the uncured prepreg with a sharp steel needle to create holes;

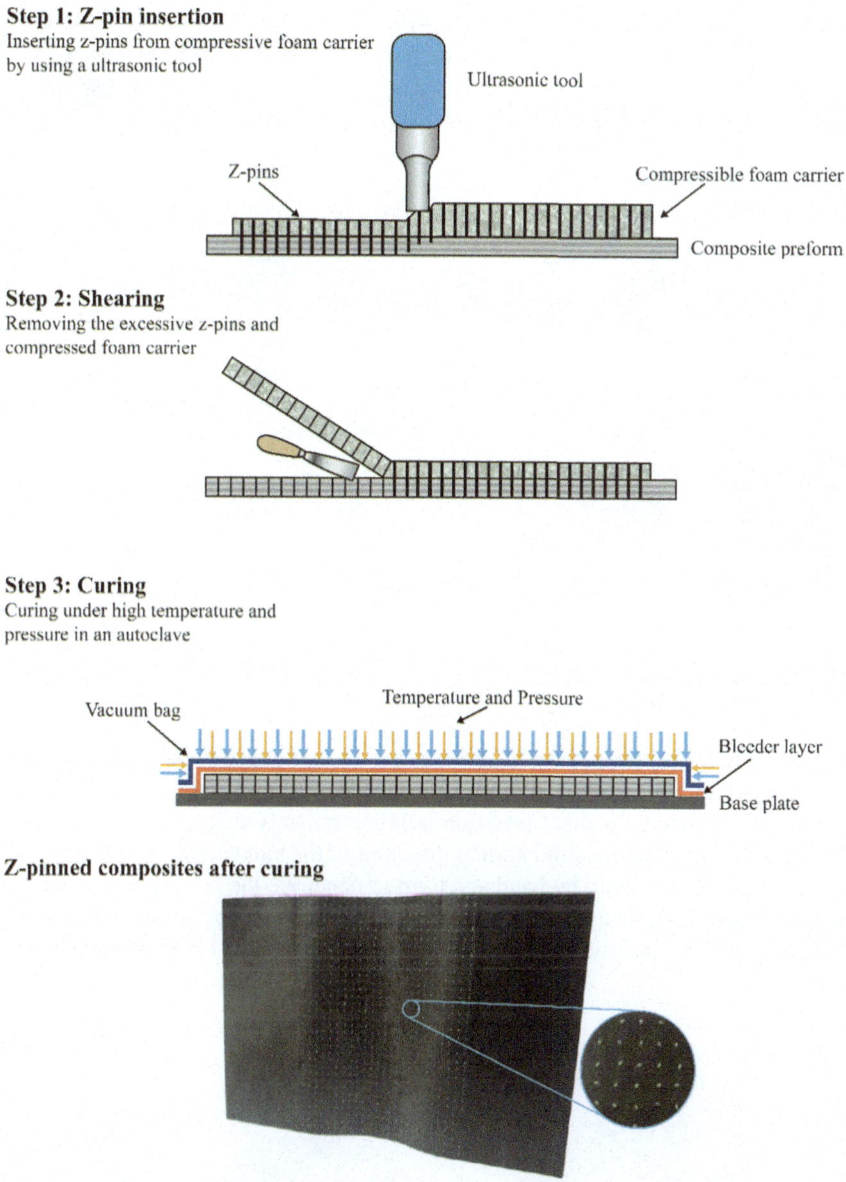


**Fig. 1.4** UAZ process for manufacturing the Z-pinned composites [28]

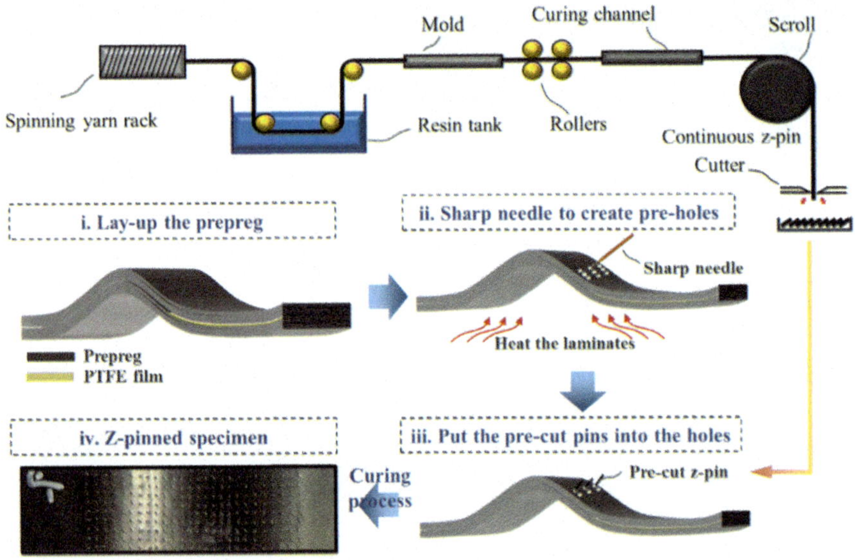


**Fig. 1.5** Schematic illustration of PHZ insertion process

(3) cut pins into the length equal to the laminates thickness;
(4) put the pins into the holes.

For the flat specimens, the PHZ process can be conducted via a universal testing machine to which a fine conical needle as attached, as shown in Fig. 1.6a. In the case of curved specimens, manual-insertion is predominantly employed to ensure that Z-pins are implanted along the normal direction of the curved region. Alternatively, a steel needle can be held by hand and used to penetrate the pre-holes manually, as shown in Fig. 1.6b. The investigation into the influence of penetration parameters reveals that a low insertion speed could significantly reduce the number of broken fibers.

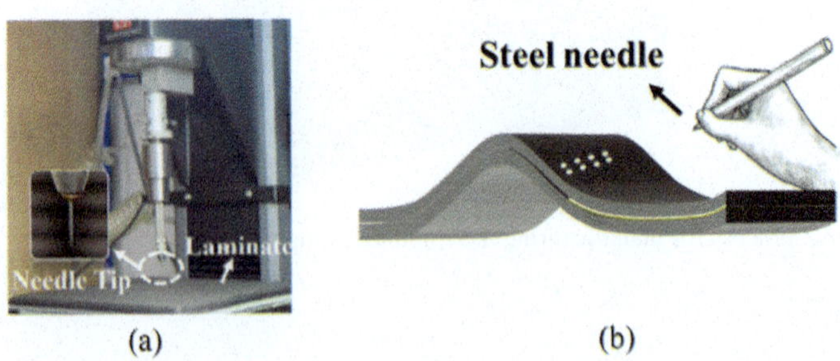


**Fig. 1.6** The insertion of PHZ: **a** a conical needle attached to a universal testing machine; **b** a steel needle held in hand

## 1.3 Microstructure of Z-pinned CFRP Composites

The insertion of Z-pins introduces microstructure changes to composite laminates including fiber waviness, ply crimp, fiber fracture, resin-rich regions, interfacial cracks and swelling. These initial defects significantly alter the mechanical properties of composite laminates, including the in-plane elasticity, strength and fatigue endurance.

In the Z-pinning process, the fibers (typically 7–15 μm) are spread laterally by the embedded Z-pins (typically 200–1000 μm), which generate local eyelet-shaped cavities and fiber waviness, as shown in Fig. 1.7. The volume fraction of affected fibers obviously depends on the shape, diameter, and volume fraction of the Z-pin. Almost 70% of fibers are affected when a volume of 4% Z-pins are inserted. It has been reported that fiber waviness adversely affects the elastic modulus and failure strength of the Z-pinned composites.

The eyelet-shaped cavities are further filled with resin during the curing process [32]. Under the elevated temperature of the curing process, the resin transforms into low viscous liquid and fills eyelet-shaped cavities before gelation. Resin-rich regions are elongated along the fiber direction and present low stiffness. At relatively dense Z-pin density ($\geq$ volume of 4%), the resin-rich regions are close enough to form the resin channel, which would further deteriorate the mechanical properties of Z-pinned composites. ~~During the insertion, the fibers right under the insertion point are bent downwards by the Z-pin tip and the affected region is limited to about 100–400 μm of the Z-pin.~~

Theoretically, Z-pins should be inserted perpendicularly to the laminates. In practice, however, Z-pins tend to deviate due to the resistance force and fiber rotation during the insertion process [33, 34]. The insertion of Z-pins also causes the swelling of laminates. For instance, laminate thickness could increase by over 15% when a volume of 4% Z-pins are inserted [35]. The laminate swelling would reduce the fiber volume fraction and thus degrade the mechanical properties of Z-pinned composites.

Another typical characteristics of Z-pinned CFRP composites microstructure is the interfacial cracks between Z-pins and host materials [36]. This is caused by

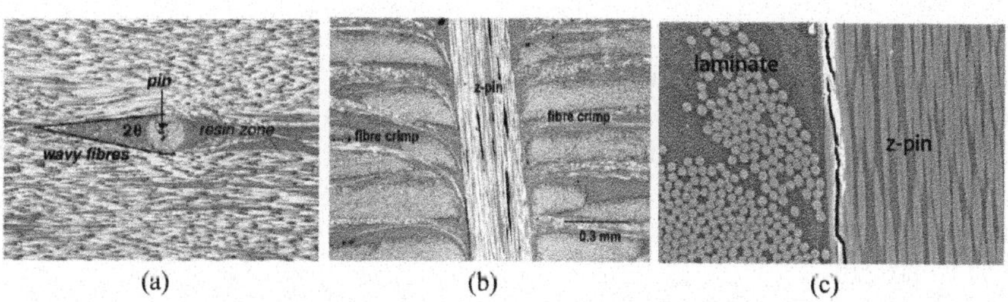


**Fig. 1.7** Typical microstructures of Z-pinned CFRP composites: **a** resin-rich zone; **b** fiber crimp; **c** initial cracks at the Z-pin/laminate interface [23]

thermal residual stresses due to different thermal expansion coefficients between Z-pins and composite laminates [1]. On the one hand, the radial residual stresses would compress the Z-pins and act as normal pressure during the pullout stage of Z-pins [7]. On the other hand, excessive axial traction probably induces partial or even complete cracking among the Z-pin/composite interfaces before mechanical loading [8, 9, 29, 34]. Hence, the Z-pin bridging mechanisms and performance changes due to the surrounding thermal residual stresses [10, 11, 31]. Finite element studies indicates that tensile thermal stresses among the Z-pin/composite interfaces could be up to four times larger than the far-field stresses and could rarely be reduced unless lower curing temperatures are applied [8].

## 1.4 Typical Applications of Z-pinned CFRP Composites

Z-pinning reinforcement stands out as a promising technical method for enhancing delamination resistance. It shows the relative simplicity and cost-effectiveness of the manufacturing process when compared to other techniques. Z-pinning have the potentials to be used as the through-thickness reinforcement of composite, core materials in sandwich panels and joint connectors in the through-thickness direction.

The C-17 Global Overlord III incorporates approximately 8% composites to reduce weight. Z-pinned sandwich structures are applied in the secondary structures and control surfaces, as depicted in Fig. 1.8a. Z-pinning is also employed in the intake ports and engine bay doors of Boeing F/A-18E/F Super Hornet to replace titanium shackles, as shown in Fig. 1.8b. This substitution results in a weight reduction of 17 kg per aircraft and a cost-saving of $83,000. Moreover, beyond the aircraft applications, Z-pinning has also found the use in reducing the weight of composite components in Formula 1 racing cars, as shown in Fig. 1.9.

## 1.5 Modeling of Z-pinned CFRP Composites

The mechanical behavior of Z-pinned CFRP composites depends on a variety of factors, such as Z-pin diameter, Z-pin volume fraction, even the processing conditions, etc. Based on a purely experimental setting, it is almost impossible to have a comprehensive understanding of the mechanical behavior of Z-pinned CFRP composites and its evolution rules under various loading conditions and processing parameters. A promising way is to develop numerical modelling techniques. Major challenges are:

(1) Influences mechanism of Z-pinning parameters on the composite properties

It is known that Z-pining is effective to increase fracture toughness and thereby improve the damage tolerance to prevent delamination extension of CFRP laminates.

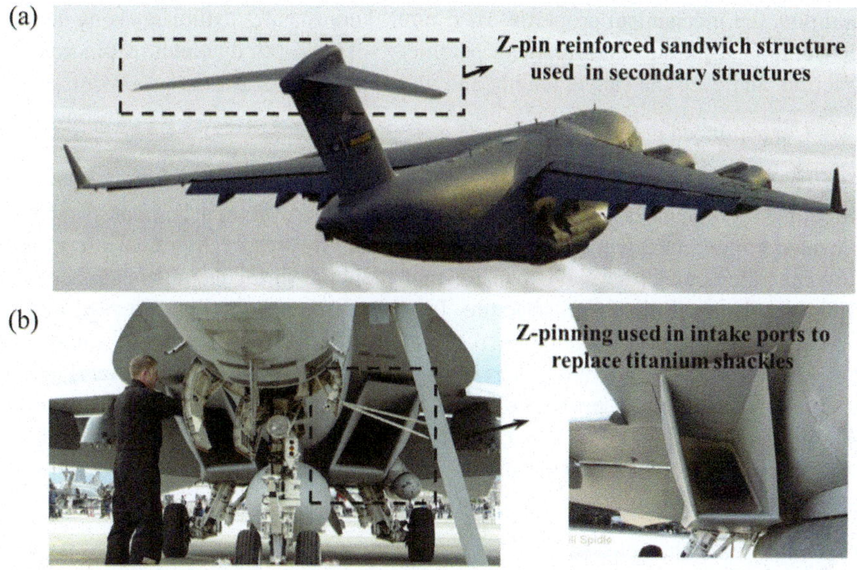


**Fig. 1.8** The application of Z-pinning in aircraft: **a** Z-pin used in C-17 Global Overlord III; **b** Z-pin used in F/A-18E/F Super Hornet

**Fig. 1.9** The application of Z-pinning in Formula 1 racing cars

When subjected to impact loading, pins can not only decrease the impact damage of the Z-pinned CFRP laminates, but also improve the residual mechanical properties. However, the Z-pining improvement mechanism is very complicated. Z-pins are usually subjected to a mix of complex loadings during the tests. The delamination resistance capacity of Z-pins is dependent on the mixed-mode ratio, due to the transition in delamination resistance mechanism and failure behavior of the Z-pins. In addition, the mechanical properties of Z-pinned composites exhibit a strong dependence upon Z-pining reinforcement parameters, i.e., Z-pin diameter, Z-pin volume fraction, and even the shape, length, and insertion angle of Z-pins.

(2) Cure-induced defects

The curing process whereafter Z-pin insertion becomes one of the essential issues affecting the enhancement of Z-pins. When Z-pinned prepreg laminates undergo elevated temperature and pressure loads in autoclave, residual stresses are gradually generated due to thermal expansion and chemical shrinkage mismatch between different phases in the microstructure. It is worth noting that the effect of curing process would not only generate residual stresses but also produce interfacial cracking around Z-pins to alter the bridging structures around Z-pins. Several in-situ morphologies evidently prove the interfacial cracking existed indeed after curing. As displayed in Fig. 1.10a, the Z-pin is partially debonded to the base laminate with interfacial cracking around it. The interfacial cracking can be also found by a side-sectional view, in Fig. 1.10b.

(3) Multiscale characteristic

Composites are essentially multi-scale with each microscopic constituent contributing to the mechanical properties of the macroscopic materials. Especially,

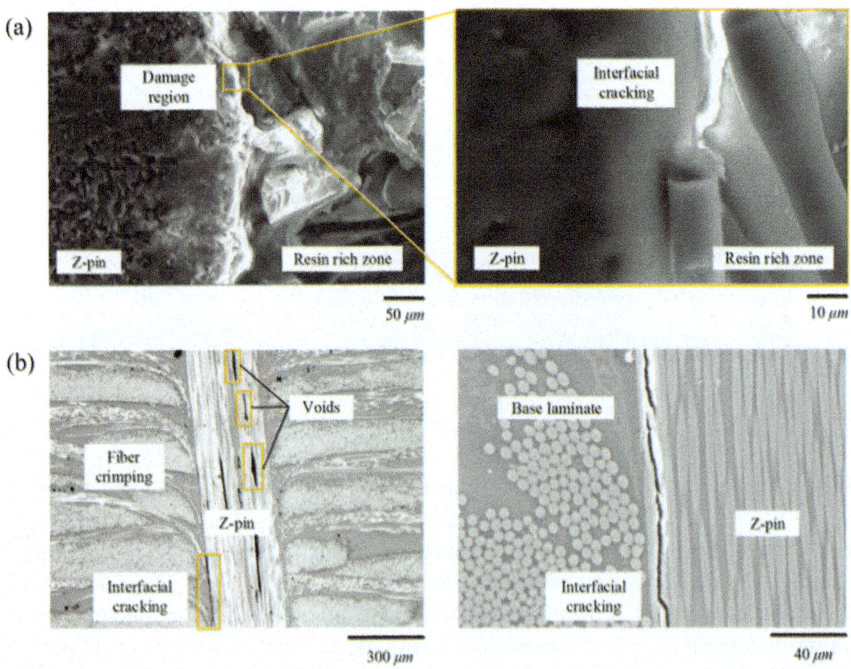


**Fig. 1.10** Partially debonding of a Z-pinned composite laminates: **a** In a cross-sectional view. **b** In a side-sectional view

the bridging and failure behavior of the individual microscale Z-pins determine the macroscopic mechanical properties of composite reinforced by numerous Z-pins subjected to various loading conditions. Discrete modelling of every single Z-pin, pin-laminate interface and surrounding laminates in detail is impractical for the macroscale cases with numerous Z-pins due to the prohibitive computation cost. Multiscale modelling is suitable to analyze Z-pinned composites. On the one hand, the behavior of individual Z-pin could be characterized in microscale simulation to provide detailed insight into the micro-mechanics and micro-structure of Z-pinned composites laminates. On the other hand, behavior of large numbers of Z-pins and composite continuum can be analyzed in macroscale simulation models where the bridging tractions of individual Z-pins are reasonably modelled.

(4) Influences of cure-induced defects on the mechanical properties

It is important to note that the presence of cure-induced defects, including residual stresses and interfacial debonding, is an inherent problem for Z-pinned CFRP composites throughout the cure. The cure-induced residual stresses and Z-pin/composite interfacial debonding affect not only the microscale bridging mechanisms of pins, but also the macroscale properties of Z-pinned composites. A deep understanding would broad the application of Z-pinning and shed light on the further synergetic optimization of Z-pinning and curing procedures.

Therefore, following issues should be addressed when designing and manufacturing Z-pinned CFRP composite products.

- What experiments needed for characterizing the mechanical properties of Z-pinned CFRP composites? How to conduct these experiments?
- How to model the evolution of residual stress and damage at resin-pin interface during the curing process of Z-pinned CFRP composites? What is the influence mechanism of Z-pinning and curing process parameters on the cure-induced defects?
- How to establish the multiscale framework to evaluate the mechanical behavior of Z-pinned composites? How to pass the microscale Z-pin bridging information to the macroscale models?
- What about the influence of cure processing conditions upon the mechanical behavior of Z-pinned CFRP composites? How to incorporate the cure processing influence into a mechanical simulation?

## 1.6 Purpose and Layout of the Book

This book attempts to develop an integrated method that combines the processing and mechanical simulations for Z-pinned CFRP composites. It can be considered as a complementary of existing books about composite mechanics and processing. To this end, the topics in this book thus cover experimental characterization, material modeling, finite element simulation as well as the integrated analysis of Z-pinned CFRP composites.

In Chap. 2, the experiment methods of characterizing the mechanical properties of Z-pinned composites are presented. Particularly, the experimental investigations over a wide range of Z-pining parameters are conducted to study the toughening effect of Z-pining on the mechanical properties of composites under mode I loading, mode II loading and low-velocity impact.

Chap. 3 is focused on the modeling of the evolution of cure-induced residual stresses and its effect on the interfacial cracking of Z-pins after the curing process. The numerical simulation results are validated against the experimental examination to justify the predictive capability of the proposed method. The effects of curing process parameters on the residual stress evolution and interfacial cracking progression are also discussed.

In Chap. 4, a multiscale framework of evaluating the delamination behavior of Z-pinned CFRP composites is presented and the effects of Z-pin/composite interfacial cracks are considered. The load–displacement relationship of Z-pinned composites under mode-I loading conditions is predicted using multiscale models and then verified experimentally. A parametric study is conducted to reveal the influence mechanism of the Z-pin/composite interfacial performance on the failure behavior of Z-pinned composites.

Chapter 5 is focused on the development of integrated computing framework that incorporates the curing defects into the mechanical simulation of Z-pinned CFRP composites. Explanations are made in detail about how to combine the cure-induced residual stress and Z-pin/composite interfacial debonding results with the finite element modeling of tensile strength, failure mode and bridging tractions.

## References


1. Bianchi F, Zhang X. A cohesive zone model for predicting delamination suppression in Z-pinned CFRP composites. Composites Science And Technology. 2011;71(16):1898–907.
2. Mouritz AP, Bannister MK, Falzon PJ, Leong KH. Review of applications for advanced three-dimensional fibre textile composite. Composites Part A: Applied Science and Manufacturing. 1999;30(12):1445–61.
3. Pingkarawat K, Mouritz AP. Improving the mode I delamination fatigue resistance of composite using Z-pins. Composites Science and Technology. 2014;92:70–6.
4. Stegschuster G, Pingkarawat K, Wendland B, Mouritz AP. Experimental determination of the mode I delamination fracture and fatigue properties of thin 3D woven composite. Composites Part A: Applied Science and Manufacturing. 2016;84:308–15.
5. Dell'Anno G, Cartié DD, Partridge IK, Rezai A. Exploring mechanical property balance in tufted carbon fabric/epoxy composite. Composites Part A: Applied Science and Manufacturing. 2007;38(11):2366–73.
6. Cartié DDR, Dell'Anno G, Poulin E, Partridge IK. 3D reinforcement of stiffener-to-skin T-joints by Z-pinning and tufting. Engineering Fracture Mechanics. 2006;73(16):2532–40.
7. Li M, Chen P. A new FE model for predicting the bridging micromechanisms of a Z-pin. Composite Structures. 2019;223:110957.
8. Sweeting RD, Thomson RS. The effect of thermal mismatch on Z-pinned laminated composite structures. Composite Structures. 2004;66(1):189–95.
9. Zhang B, Allegri G, Hallett S. Micro-Mechanical Finite Element Analysis of Single Z-pin Pull-out. In: Proceedings of SIMULIA UK Regional User Meeting. Conference, 2013.

10. Kravchenko S, Kravchenko O, Wortmann M, Pietrek M, Horst P, Pipes RB. Composite toughness enhancement with interlaminar reinforcement. Composites Part A: Applied Science and Manufacturing. 2013;54:98–106.
11. Hoffmann J, Sabban J, Scharr G. Pullout performance of circumferentially notched Z-pins in carbon fiber reinforced laminates. Composites Part A: Applied Science and Manufacturing. 2018;110:197–202.
12. Wicks SS, de Villoria RG, Wardle BL. Interlaminar and intralaminar reinforcement of composite laminates with aligned carbon nanotubes. Composites Science and Technology. 2010;70(1):20–8.
13. Srivastava VK, Gries T, Veit D, Quadflieg T, Mohr B, Kolloch M. Effect of nanomaterial on mode I and mode II interlaminar fracture toughness of woven carbon fabric reinforced polymer composite. Engineering Fracture Mechanics. 2017;180:73–86.
14. Ravindran AR, Ladani RB, Zavabeti A, Daeneke T, Wu S, Kinloch AJ, et al. Liquid metal synthesis of two-dimensional aluminium oxide platelets to reinforce epoxy composite. Composites Science and Technology. 2019;181:107708.
15. Hsieh TH, Kinloch AJ, Masania K, Sohn Lee J, Taylor AC, Sprenger S. The toughness of epoxy polymers and fibre composite modified with rubber microparticles and silica nanoparticles. Journal of Materials Science. 2010;45(5):1193–210.
16. Zeng Y, Liu H-Y, Mai Y-W, Du X-S. Improving interlaminar fracture toughness of carbon fibre/epoxy laminates by incorporation of nano-particles. Composites Part B: Engineering. 2012;43(1):90–4.
17. Mouritz AP, Jain LK. Further validation of the Jain and Mai models for interlaminar fracture of stitched composite. Composites Science and Technology. 1999;59(11):1653–62.
18. Mouritz AP, Baini C, Herszberg I. Mode I interlaminar fracture toughness properties of advanced textile fibreglass composite. Composites Part A: Applied Science and Manufacturing. 1999;30(7):859–70.
19. Fishpool DT, Rezai A, Baker D, Ogin SL, Smith PA. Interlaminar toughness characterisation of 3D woven carbon fibre composite. Plastics, Rubber and Composites. 2013;42(3):108–14.
20. Partridge I. Manufacture and Performance of Z-pinned Composites. Advanced Polymeric Composites. 2003:103–38.
21. Cartié DDR, Troulis M, Partridge IK. Delamination of Z-pinned carbon fibre reinforced laminates. Composites Science And Technology. 2006;66(6):855–61.
22. Sarantinos N, Tsantzalis S, Ucsnik S, Kostopoulos V. Review of through-the-thickness reinforced composite in joints. Composite Structures. 2019;229:111404.
23. Mouritz AP. Review of Z-pinned CFRP composites and sandwich composite. Composites Part A: Applied Science and Manufacturing. 2020;139:106128.
24. Zhang X, Chu Q, Xiao J. Experimental study on the performance of twisted fiber reinforced composite Z-pin. 21st International Conference on Composite Materials. 2017.
25. Wang XX, Chen L, Jiao YN, Li JL. Experimental study on interfacial adhesive properties between twisted Z-pin and laminates. Guti Huojian Jishu/Journal of Solid Rocket Technology. 2014;37:856–62.
26. Wang X, Chen L, Jiao Y, Li J. Preparation of polyvinyl alcohol-matrix Z-pins and tests of their piercing properties. Journal of Reinforced Plastics and Composites. 2014;33(24):2255–64.
27. Boyce JSN, Freitas GAF, Magee CLW, Fusco TMM, Harris JJN, Kunkel EN. Ultrasonic fastening system and method. 1998.
28. Cheng J, Xu Y, Zhang W, Liu W. Multiscale modelling of curing-induced Z-pin/composite interfacial cracks. International Journal of Mechanical Sciences. 2023;240:107924.
29. Chang P, Mouritz AP, Cox BN. Properties and failure mechanisms of Z pinned CFRP composites in monotonic and cyclic tension. Composites Part A: Applied Science and Manufacturing. 2006;37(10):1501–13.
30. Yasaee M, Lander JK, Allegri G, Hallett SR. Experimental characterisation of mixed mode traction–displacement relationships for a single carbon composite Z-pin. Composites Science and Technology. 2014;94:123–31.

31. Hoffmann J, Scharr G. Pull-out performance of rectangular Z-pins in hot-cured carbon fiber reinforced laminates. Composite Structures. 2018;186:62–7.
32. Cui H, Li Y, Koussios S, Zu L, Beukers A. Bridging micromechanisms of Z-pin in mixed mode delamination. Composite Structures. 2011;93(11):2685–95.
33. Ratcliffe J, O'Brien T. Discrete Spring Model for Predicting Delamination Growth in Z-Fiber Reinforced DCB Specimens. In: Proceedings of ASC/ASTM-D30 Joint 19th Annual Technical Conference. 2004.
34. Zhang B, Allegri G, Yasaee M, Hallett SR. Micro-mechanical finite element analysis of Z-pins under mixed-mode loading. Composites Part A: Applied Science and Manufacturing. 2015;78:424–35.
35. Chang P, Mouritz AP, Cox BN. Properties and failure mechanisms of pinned composite lap joints in monotonic and cyclic tension. Composites Science and Technology. 2006;66(13):2163–76.
36. Mouritz AP, Koh TM. Re-evaluation of mode I bridging traction modelling for Z-pinned CFRP composites based on experimental analysis. Composites Part B: Engineering. 2014;56:797–807.

# Chapter 2
# Experimental Tests of the Mechanical Properties

The mechanical properties of Z-pinned composites exhibit a strong dependence upon Z-pining parameters, i.e., Z-pin diameter, Z-pin volume fraction. The toughening effect of Z-pin on interlaminar property directly determines the delamination resistance of composite under impact loading. To characterize the mechanical properties of Z-pinned CFRP composites, the most common method is based on the experimental tests, including low-velocity impact test to explore the damage tolerance after impact loading, double cantilever beam (DCB) test for mode I property and end notched flexure (ENF) test for mode II property, as shown in Fig. 2.1.

This chapter focuses on the experimental tests to investigate the toughening effect of Z-pining on the composites under mode I loading, mode II loading and low-velocity impact. To reduce the initial damage induced by Z-pin insertion process, the manufacturing of all the specimens is based on the low-damage pre-hole Z-pinning (PHZ) technique. Most of the Z-pinned composite structures in engineering applications have curved shapes, for instance the composite turbine blade. Thus, curved specimens are designed and discussed in this chapter.

## 2.1 Mode I Delamination Resistance

### *2.1.1 Materials Preparation and Experimental Setup*

The specimen is fabricated with carbon fiber epoxy prepregs T700/7901 supplied by Weihai Guangwei Composites Co., Ltd. The double cantilever beam (DCB) experiment is carried out in accordance with ASTM D5528 standard. The unidirectional prepreg plies are stacked with a ply scheme of $[45/0/-45/90]_{4s}$ and the prepreg is debulked every four plies to vacuum air and compact the laminates. The DCB specimen used in mode I tests has a dimension of 150 mm × 25 mm, as shown in Fig. 2.2, which is designed based on the geometrical model of a composite turbine



Y. Xu and W. Zhang, *Modeling of Z-pinned Carbon Fiber-Reinforced Polymer (CFRP) Composite*, https://doi.org/10.1007/978-981-96-3628-0_2

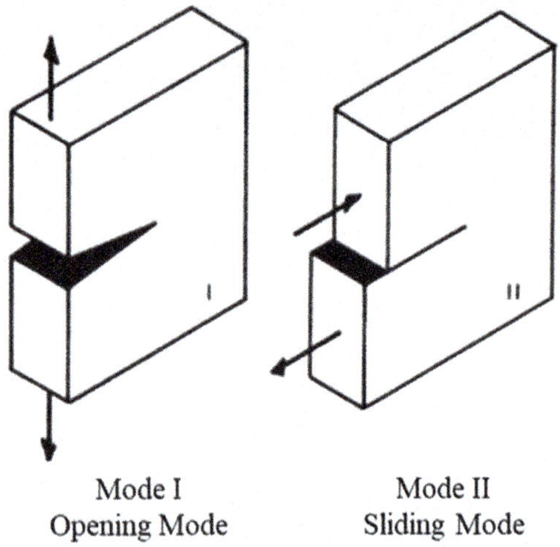


**Fig. 2.1** Two typical fracture modes

blade. To ensure perpendicular load on to the surface during the testing, a pair of metal piano hinges are bonded to both ends of the specimen faces. The L-shaped hinges are bonded using epoxy adhesive, and the bonding surfaces are sanded with 280 grit sandpaper before adhesion. At one end of each specimen, a PTFE film with a length of 54 mm and a thickness of 12 μm is present to initiate a crack at the mid-plane of the laminate during the layering process.

The solid cylindrical Z-pins are produced by pulling continuous polyimide fiber filaments through resin and then pultruding them through a circular mold, as shown in Fig. 2.3. The twisting process could squeeze excess bubbles in the fiber bundle, thereby reduce initial defects. Besides, the twisted pins increase the contact area with

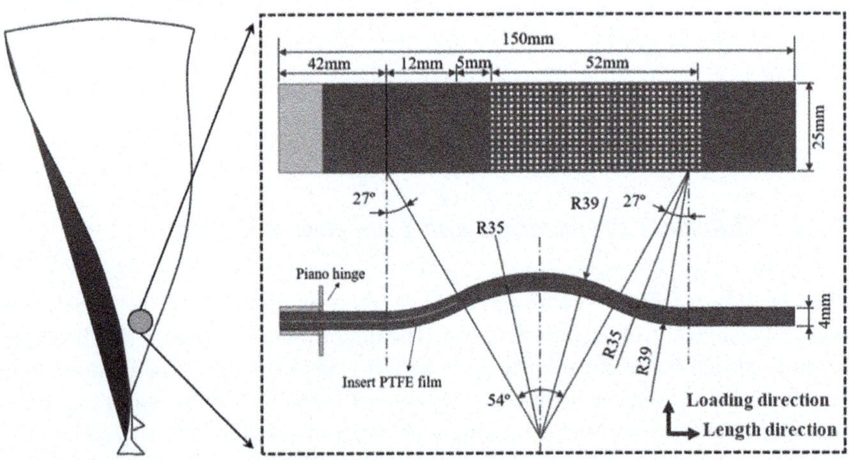


**Fig. 2.2** Geometry and dimensions of the DCB specimen

the laminate. In the test, the twist degree is 80 twist per meter. The Z-pinned region of the specimens starts 5 mm behind the pre-crack tip and covers an area of 50 mm × 25 mm. The PHZ process, as illustrated in Fig. 2.4 is applied to insert the pins into the uncured laminates. Firstly, a stainless needle is used to penetrate the uncured laminates with the support of a designated apparatus to ensure the verticality between the holes and the curved surface. The diameter of the needle varies according to the Z-pin diameter. The holes are uniformly distributed in the Z-pinned region. Then, the pins with the same length as the thickness of the laminate are manually inserted into these holes. Finally, all the specimens are cured in an autoclave for 4 h at an overpressure of 0.6 MPa and a temperature of 120 °C. It is noticed that the prepreg is heated properly in advance to ease the insertion of Z-pins within the softened resin matrix. The specimens with thin pin (0.3 mm diameter) and thick pin (0.5 mm diameter) are marked with "a" and "A", respectively. The number followed by "a" and "A" represents the implant spacing of the pins.

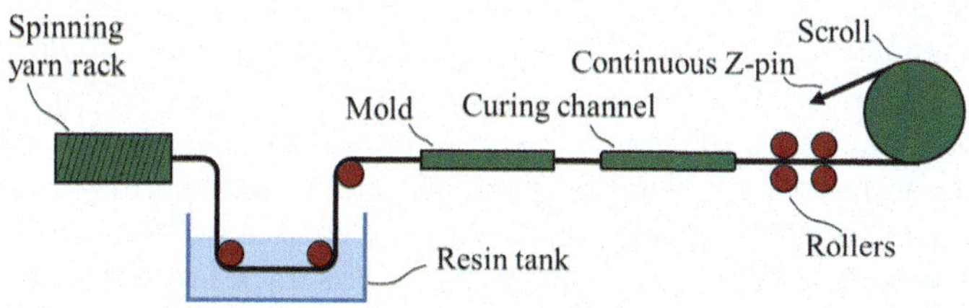


**Fig. 2.3** Schematic diagram of twisted continuous Z-pin pultrusion process

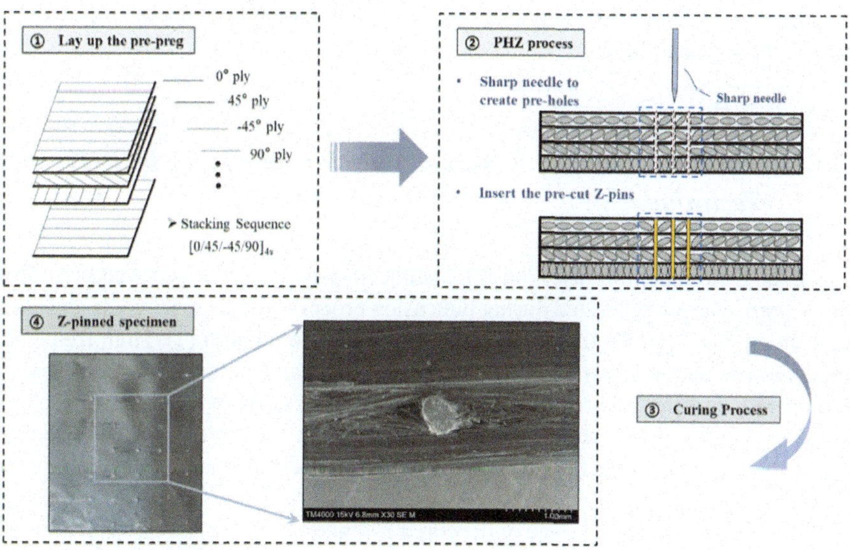


**Fig. 2.4** Process of pre-hole Z-pin insertion technique

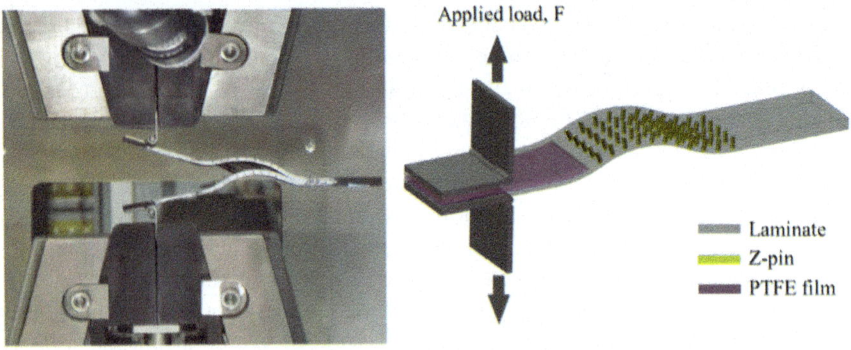


**Fig. 2.5** Schematic illustration of the DCB test process

The DCB test is conducted using a DNS-100 universal machine with load capacity of 100 kN, as shown in Fig. 2.5. To facilitate visualization, one edge of the DCB specimen is marked in write and labeled with a ruler for sharp contrast. A LED light is positioned next to the specimen to make it easier to detect the crack tip. The determination of fracture toughness is performed using monotonic displacement control at a constant crack opening displacement rate of 1 mm/min to guarantee steady crack propagation. As the opening displacement increases, delamination propagates along the mid-plane of the specimen from the pre-crack in short increments (1–10 mm). At each increment, the applied load (F), crack opening displacement (δ), and crack length (a) are measured and recorded. The motion of the delamination front is monitored with a high-definition electron microscope, which provides an accuracy of ± 1 mm. For each testing group, three parallel specimens are tested to ensure the repeatability.

### *2.1.2 Microstructure of Z-pinned CFRP Composites by PHZ Technique*

For UAZ technique, the insertion force subjecting to Z-pins is generated by ultrasonic horn. The air generated during pultrusion process would create voids and elongate along the Z-pin length under the action of ultrasonic horn [1]. High frequency compressive waves are transmitted into the Z-pin carriers, and then the pins are hammered into the prepreg. This process generally leads to the fiber crimp and breakage [2]. For PHZ technique, a steel needle with a sharp tip is used to replace Z-pin to create holes. In this way, Z-pin is well preserved and away from internal fiber damage.

Here, the microscopic image of the cured Z-pinned specimen is shown in Fig. 2.6. It can be seen that the Z-pin under the PHZ process is fairly straight with a very small

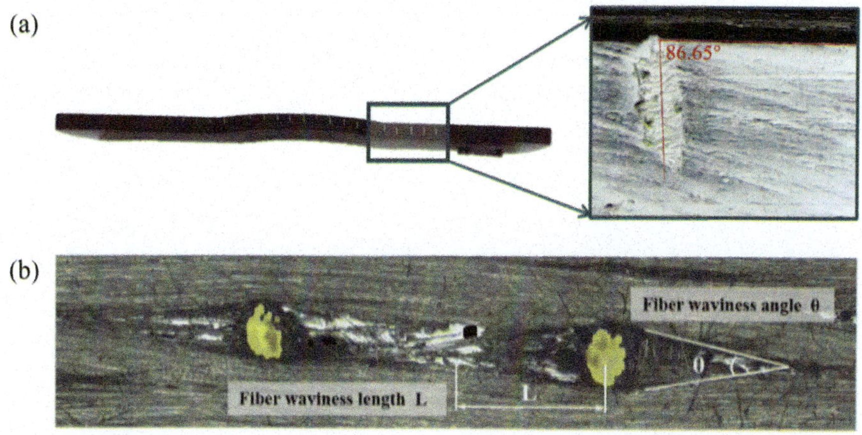


**Fig. 2.6** Microscopic image of cured Z-pinned specimen under the PHZ process: **a** offset angle of Z-pin; **b** fiber waviness length and angle

misalignment angle of 3.35°, while the UAZ process [2] results in a misalignment angle between13.8° and 23.4°.

It is observed that eye-shape resin-rich zones are developed and elongated along the fiber direction of the laminate. Fiber waviness is induced on both sides of the resin-rich zone. Fibers are spread laterally when they pass around the Z-pin, and thus the flowing resin can be full of the eye-shape gap. The values of the length and angle of fiber waviness shown in Fig. 2.6b are recorded for specimens with thin (a5) and thick (A5) Z-pins. For each specimen, 6 groups of data are randomly selected and presented in Fig. 2.7. It can be seen that the fiber waviness length and angle of specimen with thick Z-pin are slightly higher than those with thin Z-pin. For specimen with Z-pin of 0.3 mm, the average length and angle of fiber waviness are 1.07 mm and 6.5°. For specimen with Z-pin of 0.4 mm, they are 1.4 mm and 7.2°. Generally, the fiber waviness length and angle mainly depend on the diameter of Z-pin and reflect the size of rein-rich zone. They are increased both along with the increase of Z-pin diameter. Comparatively, fiber waviness length and angle are 1 mm and 8.5° for UAZ specimen with Z-pin of 0.28 mm [2].

### *2.1.3 Data Reduction Scheme for Curved Specimens*

#### Z-pin Misalignment Definition

As the mode I displacement increases during the DCB test process, the Z-pins located at various positions undergo varying sliding displacements, leading to differences in mode-mixity among the pins within a specimen due to the curvature. Here a simplified method is applied to define the misalignment of Z-pins at different positions. The

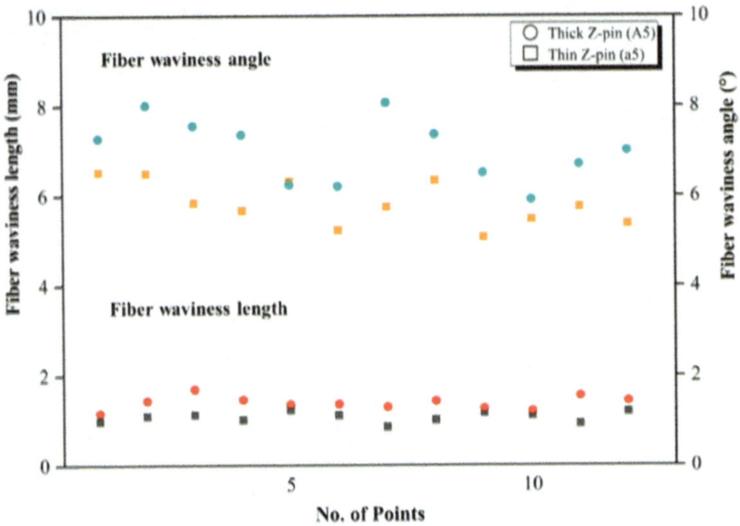


**Fig. 2.7** Fiber waviness length and angle for PHZ specimens with thin (a5) and thick (A5) Z-pins

mode-mixity is defined by the direction of the load with respect to the pin axis, as illustrated in Fig. 2.8. Due to the curvature of the specimens, the angle between Z-pin axis direction and the loading direction ranges from approximately 0° to 25°. Additionally, based on the measurement in Fig. 2.6, the Z-pin misalignment with the through-thickness axis of the laminates is averagely determined as 3.8°. The deviation of each individual pin in the $X$ and $Y$ directions is measured and considered in the calculation of mode-mixity ($\phi$), in Eq. (2.1). This method was introduced by Yasaee et al. [3, 4].

$$\phi = \sqrt{\cos^2\chi\sin^2\zeta + \sin^2\chi(\sin^2\psi + \cos^2\psi\cos^2\zeta) - \frac{1}{2}\sin 2\chi\sin 2\zeta\cos\psi} \tag{2.1}$$

where $\chi$ is the angle between Z-pin axis and the loading direction. $\zeta$ and $\psi$ represent the offset angles of the pins with the $X$- and $Z$-axis respectively, both values can be calculated in terms of the misalignment angles $\alpha_{13}$ and $\alpha_{23}$ by Eqs. (2.2) and (2.3):

$$\tan\zeta = \sqrt{\tan^2\alpha_{13} + \tan^2\alpha_{23}} \tag{2.2}$$

$$\tan\psi = \frac{\tan^2\alpha_{23}}{\tan^2\alpha_{13}} \tag{2.3}$$

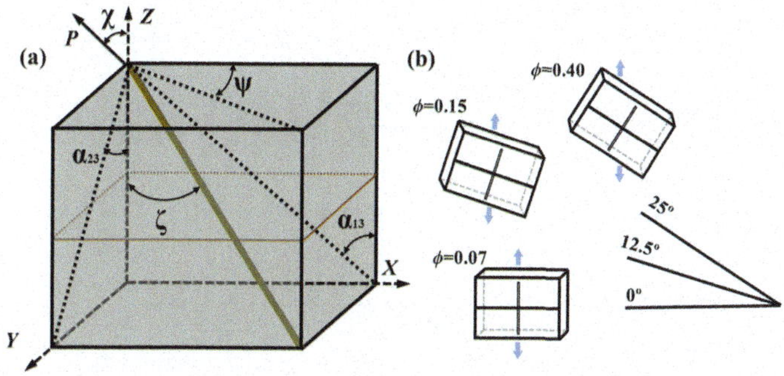


**Fig. 2.8** **a** A single Z-pin misalignment utilized for determining the mode-mixity in relation to the loading direction; **b** illustration of varying angles of Z-pins located at different positions along a curved specimen

### Comparation of Timoshenko Beam Theory and Euler–Bernoulli Beam Theory

The interlaminar strain energy release rate is an important parameter for characterizing the Mode I delamination resistance in DCB test. To determine the interlaminar strain energy release rate, the compliance of the specimen should be firstly obtained. In general, Timoshenko beam theory and Euler–Bernoulli beam theory can be used to calculate the compliance of the flat and curved specimens, respectively. The compliance of Timoshenko beam theory, is expressed by Eq. (2.4) [5]:

$$C_{\text{flat}} = \frac{8}{bE_\theta}\left(\frac{a}{h}\right)^3 + \frac{12}{5bG_{\theta z}}\left(\frac{a}{h}\right) \tag{2.4}$$

where $h$ is the half thickness of the specimen, $a$ is crack length, $b$ is specimen width, $E_\theta$ is tensile modulus in fiber direction, $G_{\theta z}$ is shear modulus. Note that in Euler–Bernoulli beam theory, the shear effect is omitted by eliminating the second term in Eq. (2.4). The magnitude of the shear effect depends on $G_{\theta z}$ and the ratio of $a$ and $h$ [6]. For a Timoshenko beam, the compliance can be further divided into two segments, as expressed in Eq. (2.5) [5]:

$$C_{\text{curved}} = (C_{\text{curved}})_{\text{top}} + (C_{\text{curved}})_{\text{bottom}} \tag{2.5}$$

with

$$\begin{aligned}(C_{\text{curved}})_{\text{top}} &= \frac{1}{2}\frac{H^+}{E_\theta}\left(\cos\frac{a^+}{R}\sin\frac{a^+}{R} - \frac{a^+}{R}\right)\\ &\quad + \frac{6}{5b}\frac{R/h + 1/2}{G_{\theta z}}\cos\frac{a^+}{R}\sin\frac{a^+}{R}\end{aligned} \tag{2.6}$$

$$(C_{\text{curved}})_{\text{bottom}} = \frac{1}{2}\frac{H^-}{E_\theta}(\cos\frac{a^-}{R}\sin\frac{a^-}{R} - \frac{a^-}{R}) + \frac{6}{5b}\frac{R/h - 1/2}{G_{\theta z}}\cos\frac{a^-}{R}\sin\frac{a^-}{R} \quad (2.7)$$

where, R is the radius of the curved specimen, and

$$H^+ = \frac{R/h + 1/2}{b - b(R/h + 1/2)\log(\frac{R/h+1}{R/h})}$$

$$\frac{a^+}{R} = \frac{R/h + 1/2}{R/h}\frac{a}{R}$$

$$H^- = \frac{R/h + 1/2}{b - b(R/h + 1/2)\log(\frac{R/h}{R/h-1})}$$

$$\frac{a^-}{R} = \frac{R/h - 1/2}{R/h}\frac{a}{R}$$

The difference between the curved beam theory (Euler–Bernoulli beam theory) and flat beam theory (Timoshenko beam theory) is analyzed by plotting the compliance versus *a/h* (the crack length to thickness ratio) in Fig. 2.9. The deviation between the two theories increases as *R/h* decreases and *a/h* increases. For the case where *R/h* = 18 and $E_\theta/G_{\theta z} = 29$, the deviation from the curved beam theory is over 20%. The flat beam theory posits an infinite shear stiffness for beams, leading to pure bending deformation with no shear strain in the cross section. On the other hand, the curved beam theory considers the shear modulus of the material, causing shear strain in the cross section when the beam is curved and deformed. This results in a relatively minor difference between the two theories for long and thin beams. However, as *R/h* decreases and *a/h* increases (crack propagation), the beam gradually transforms into a short and thick beam, and the impact of shear modulus becomes significant. This causes a substantial increase in the deviation between the flat beam theory and the curved beam theory. Moreover, for materials with a high $E_\theta/G_{\theta z}$ ratio (equal to 29), the assumption of infinite shear stiffness is invalid, leading to a more conspicuous divergence between the two beam theories. Therefore, the curved beam theory is used for the calculation of compliance in this study. The following analysis of strain energy release rate of unpinned and Z-pinned specimens are both based on the curved beam theory.

**Data Reduction Scheme**

In this study, the interlaminar strain energy release rate $G_c$ is calculated as:

$$G_C = G_{IC} + G_{IIC} \quad (2.8)$$

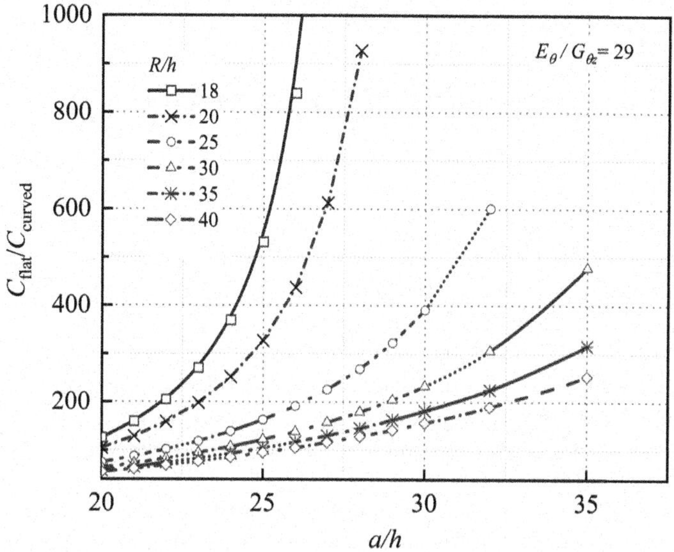


**Fig. 2.9** Compliance ratio (flat beam theory to curved beam theory) versus a/h (crack length to thickness ratio)

where $G_{IC}$ and $G_{IIC}$ denote the mode I and mode II strain energy release rates, respectively. Both values are calculated as [7]:

$$G_{IC} = \frac{12F^2}{E_\theta b^2 h}\left(\frac{ac}{h}\right)^2 \tag{2.9}$$

$$G_{IIC} = \frac{9F^2}{4E_\theta b^2 h}\left(\frac{a_c}{h}\right)^2 \tag{2.10}$$

where $F$ is applied force; $a_c$ is modified crack length, $a_c = |a + \Delta|$, $\Delta$ represents the relation between the compliance and the crack length of specimen. $\Delta$ can be obtained by fitting the compliance ($C$) with the initial crack length ($a$); $b$, $h$, are width, and thickness of the specimen, respectively.

Unpinned specimens exhibit smaller in-plane fiber bridging in the crack area, leading to a delamination prevention mechanism termed as small-scale bridging. In contrast, for Z-pinned specimen, the failure procedure generates a bridging area that is of the order of magnitude of the specimen width, creating a large-scale bridging zone. The fracture toughness calculated for these specimens is regarded as "apparent fracture toughness" due to the presence of reinforcement.

## 2.1.4 Experimental Results and Analysis

### Fracture Toughness of Z-pinned and Unpinned Specimens

The representative force–displacement curves for unpinned and Z-pinned specimens under different parameter including Z-pin diameter and Z-pin volume fraction are displayed in Fig. 2.10a. It is observed that all the specimens experience an initial linear increase before a slight load drop from the pre-crack, which indicates the onset of stable crack growth. After the load drop, the trends between unpinned and Z-pinned specimens are significantly different.

For the unpinned specimens, a sudden load drop occurs after reaching the maximum load of 105 N due to the reduction in equivalent bending rigidity caused by the specimen curvature. The lack of interlaminar toughening in unpinned specimens leads to a few instances of stick–slip behavior and early failure due to the reduction in bending rigidity [6].

The overall maximum load of Z-pinned specimens, which increase more significant than that of unpinned specimens, directly corresponds to the toughening parameters. At the load increase stage, oscillations in the fracture force indicate the involvement of delamination propagation around the pins. The load increases sharply until Z-pin bridging area is fully generated. When the cracks arrive at the reinforced area, Z-pins are triggered and transfer axial stresses to the surrounding laminates to reduce the stress concentration on the crack tip and create bridging traction on the upper and lower surface. The first pull-out of Z-pin coincides well with the overall peak force.

As the cracks pass through pins, Z-pins are prone to debonding from the laminates, resulting in interfacial cracks between pin and resin matrix. As the displacement increases, Z-pin rows are pulled out from the laminates, leading to unstable crack propagation and stick–slip behavior observed in the force–displacement curves. This phenomenon is dependent on the ratio of debonding energy and frictional work operating across the debonding face [8]. This is indicated by the "steps" in Fig. 2.10a.

Once the bridging zone is fully developed, the involvement and failure of Z-pin form a dynamic balance. The length of the bridging zone remains constant with increasing delamination and the bridging area moves forward as the crack propagation. For specimens with relative higher fraction of pins, Z-pins can more effectively retard crack propagation, as evidenced by the higher load and longer displacement in Fig. 2.10b. Once the Z-pins located at the crack tip are pulled out, the accumulated energy is suddenly released, leading to quick crack propagation to the next row of Z-pins. In the DCB tests process, this results in one or more periods of no crack propagates followed by speedy propagation, which can be characterized by a sharp load drop in the force–displacement curves. One of the main advantages of Z-pinning is that the load force for delamination does not drop completely to arbitrarily small values as the crack propagation, providing a guarantee for the safety of the structure in engineering application. In some cases, once the number of Z-pins bridging the gap between the cantilevers is reduced to a few, a sudden opening of the specimen is observed. This is illustrated by a sudden load drop in the yellow curve in Fig. 2.10a.

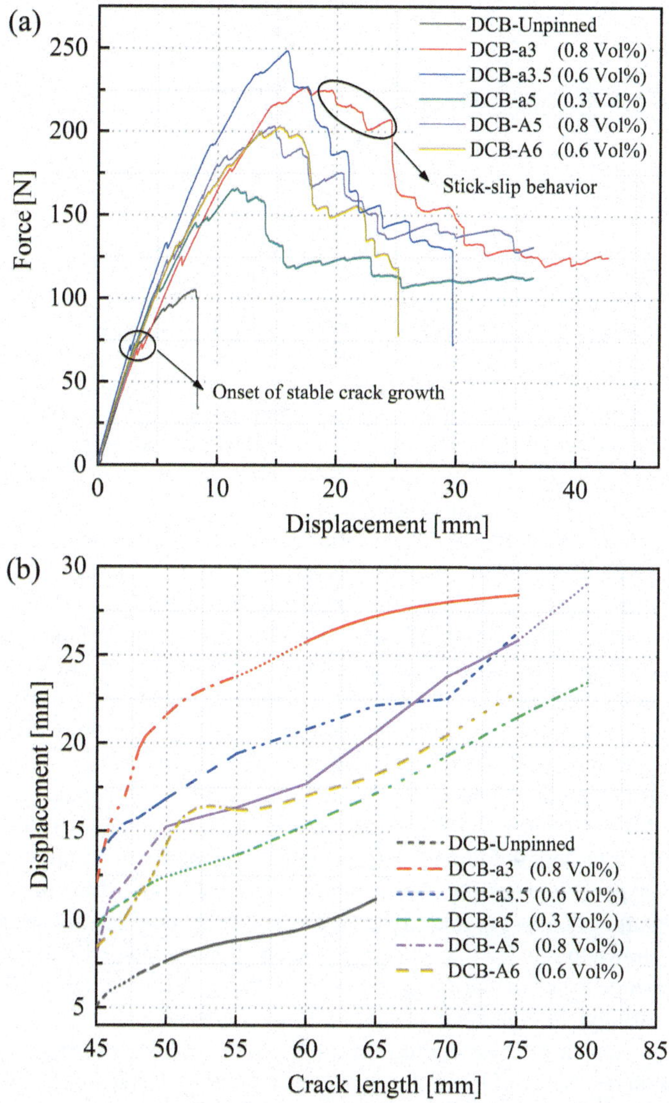


**Fig. 2.10** **a** Representative force–displacement curves; **b** representative displacement-crack length curves

After the DCB test, only a small number of Z-pins are observed to rupture, which is believed to be caused by localized bending of Z-pins due to the shear stress.

The increasing in fracture toughness with crack propagation is drawn by resistance curves (R-curves). The disparity in R-curves of the unpinned and Z-pinned specimens is influenced by the additional bridging force provided by pins, which varies depending on the reinforcement parameters. Both unpinned and Z-pinned CFRP

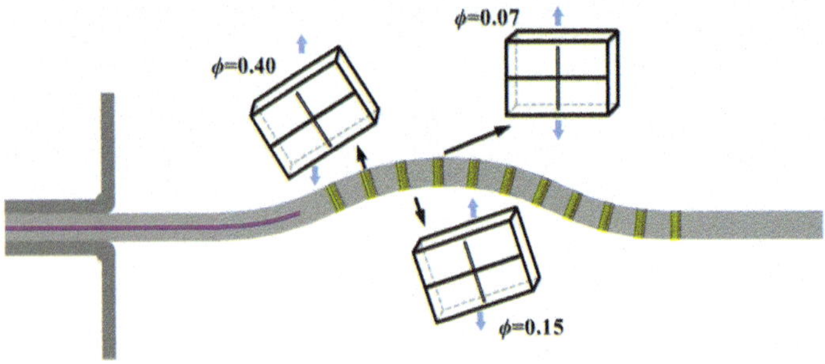


**Fig. 2.11** Schematics of the variation of $G_I$-to-$G_{II}$ ratio at different locations

composites exhibit scatter in the fracture toughness data. The scatter in unpinned specimens is caused by stick–slip process induced by fiber bridging along the crack surface. Additionally, it is observed that the crack growth rate is increased when Z-pins at the rear of the bridging area failed, while the rate is slowed when the crack tips arrive at a new row of pins. The scatter in Z-pinned CFRP composites is more significant, which can be attributed to the variability in deflection angles and mode-mixity of Z-pins at different position along the curved laminates (Fig. 2.11). The scattering of data in both laminates can be attributed partly to the challenges faced in precisely identifying the location of the crack tip during the testing, despite the usage of an optical microscope. It is also found that a relative stable fracture toughness can be observed when the pins are progressively pulled out row by row. Conversely, the apparent fluctuation in $G_c$ corresponds the fracture of the pins.

The fracture toughness value required to initiate cracks is dependent on the crack length for both unpinned and Z-pinned CFRP composites. The $G_c$ value increases slightly with crack length for unpinned specimens, but increases rapidly for Z-pinned laminated, especially when the Z-pin volume fraction is high. Based on the analysis above, the ratio of $G_I$-to-$G_{II}$ initially increases and then decreases as the crack propagates, as shown in Fig. 2.11. The fracture toughness for unpinned specimens shows a weak dependence of the $G_I$-to-$G_{II}$ ratio. Conversely, the Z-pinned CFRP composites demonstrated a heightened sensitivity to the $G_I$-to-$G_{II}$ ratio. The strengthened sensitive of Z-pinned CFRP composites to the mixed-mode ratio suggests that the delamination resistance of the Z-pins is more sensitive to the $G_I$-to-$G_{II}$ ratio than the composite materials itself.

Unpinned specimens exhibit a relative smoother fracture toughness compared to reinforced ones, as the energy is dissipated by the formation of the crack surface, determined by the mechanical properties of the matrix. The significant improvement in fracture toughness of Z-pinned specimens is due to the large-bridging zone along the delamination crack. The crack bridging mechanism and failure mode of Z-pin along the cracks surface are primarily governed by mode-I loading, including debonding and pulling out, viewed as energy absorption. This dominance of the results is partially due to the energy-consuming capacity of the resin itself is considerably weaker than of Z-pin debonding and pull-out.

In addition, the capacity of the pin/laminate interface to deliver the bridging force during pull-out process is crucial for the delamination resistance performance of reinforced laminates. It has been observed that for Z-pinned specimens under the same volume fraction, laminates with thinner Z-pins (0.3 mm) exhibit higher fracture toughness compared to the those with thicker pins (0.5 mm). This is attributed to the longer Z-pin/laminate interface contact area provide by the thinner pins at the same volume fraction, resulting in a larger bonding area and improved the maximum debonding force and friction. This then results in increased shear stresses within the interface and an increase in bridging force during pull-out, as depicted in Fig. 2.12b.

The formation of resin-rich zone during the PHZ process leads to the generation of two contact regions: resin matrix and fiber, which have different interaction with Z-pin. During the curing process, the thermal expansion coefficients of pin and laminates differ, resulting in debonding at the interface [9]. For the thicker pins, larger contact area with the matrix and higher stress reduces the maximum debonding force. On the other hand, debonding is more likely to occur around thicker pin and extend in the thickness direction. This can partly explain why the specimens with thinner pins exhibit higher fracture toughness.

### Failure Mechanism of Z-pinned CFRP Composites

Fractographic examination of the Z-pinned CFRP composites using electron microscopy reveals the crack bridging mechanism and failure mode of Z-pins in delamination propagation process, as shown in Figs. 2.13 and 2.14. As the double cantilever beams of the DCB specimens separate from each other, pins are progressively pulled out from the surrounding laminates via interfacial shear failure, and it is noticed that the lower sections of the PHZ insertion process are especially susceptible to pull-out. This is attributed to the limited interfacial bonding between the pin and the laminate. During the pull-out process, the robust interaction between Z-pins and laminates and the significant frictional resistance against Z-pin pull-out are the primary factors contributing to the enhanced fracture toughness of Z-pinned CFRP composites. When the pins are subjected to high longitudinal-transverse shear stresses, they are split into strands and slide relative and prove further loading before failure.

At the initial stage of Z-pin failure, fiber bridging behavior is observed around pins. As the pins are pulled out, the frictional force between Z-pin and laminates induces a great number of interlaminar micro-cracks, which helps to improve the fracture toughness (the $G_c$ value). It is observed that Z-pins cause crack branching as the delamination bypassed them (as shown in Fig. 2.13b), which effectively retards the growth rate of delamination cracks at this stage.

Reinforced specimens require higher displacement to form cracks of the same length compared to unpinned specimens, which implies that pins under higher volume fraction undergo more severe bending moment during pull-out. In the case in which bending Z-pins (for the part embedded in the laminates) plough the laminate, the

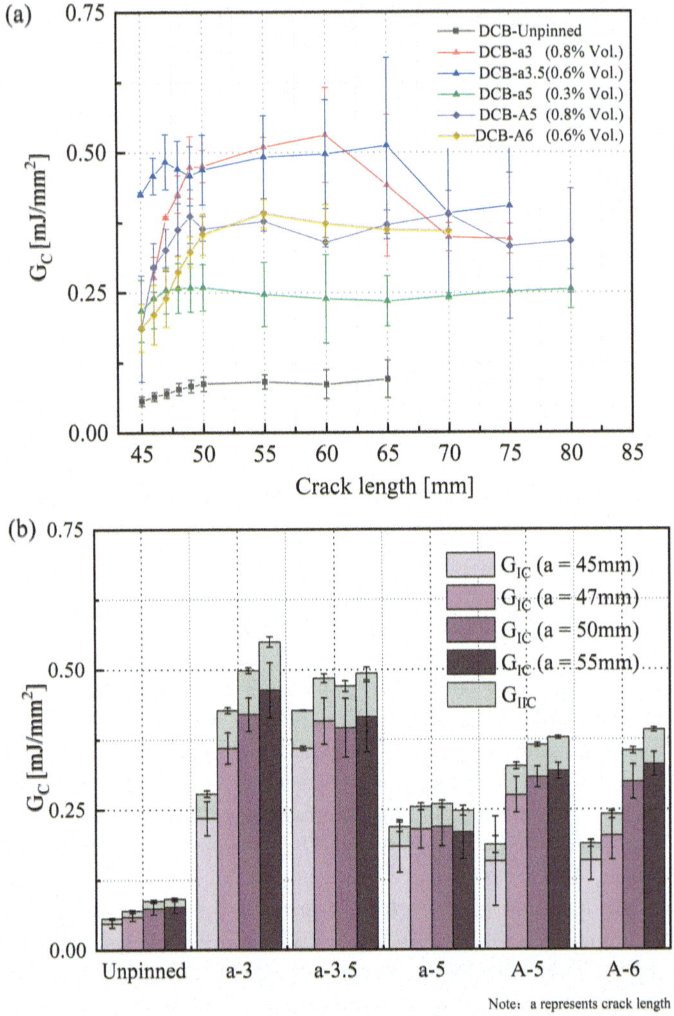


**Fig. 2.12** **a** Representative delamination resistance curves; **b** effect of the Z-pin parameter on the interlaminar fracture toughness $G_C$

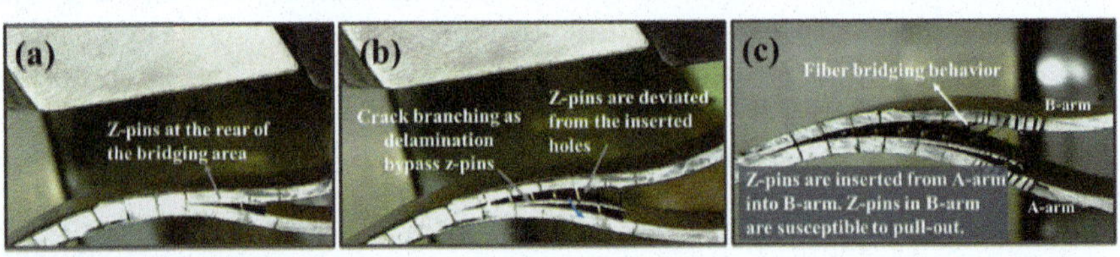


**Fig. 2.13** Captures of crack propagation stage of Z-pinned laminate **a** cracks initiate at Z-pinned region; **b** crack extension; **c** Z-pin pull-out

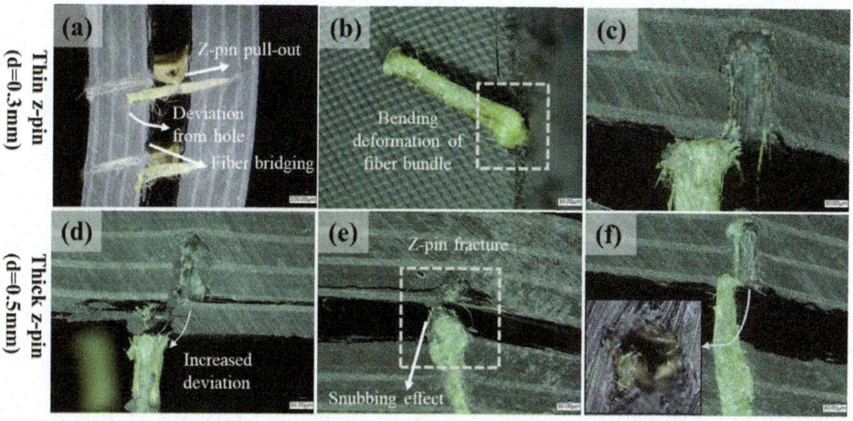


**Fig. 2.14** Optical photograph of Z-pin after DCB test; **a** deviation from holes; **b** buckling of Z-pin fiber; **c** Z-pin pull-out **d** increased deviation; **e** fiber fracture **f** residues in holes

bridging force remains constant at a high value to resist pull-out over a long displacement. As crack propagate, the pins at crack tip endure the highest traction force. The deflection of the curved beam and the orthotropic property of Z-pin cause fiber bulking of Z-pin at the midplane of the laminates, as shown in Fig. 2.14b. With the increasing of shear stress, Z-pins are locally deformed at the delamination crack plane and eventually fractured, as captured in Fig. 2.14e. During the failure process of pins, there is a traction force component perpendicular to the fiber direction, with the maximum transverse force component occurring at the fiber bulking area. The significant transverse shear force that Z-pin experiences in curved arms becomes a critical factor in the failure mode of the Z-pin during the process of arm opening. In addition, the transition in delamination resistance mechanism and failure behavior of Z-pins, from combined tensile-shear fracture during mixed-mode loading to pull-out during mode-I dominant loading, clarifies why the delamination resistance capacity of pins is dependent on the mixed-mode ratio.

After the post-test observations of fractured Z-pins ends, it has been discovered that pins are protruded from the fracture surface in many cases, and the gouging of fibers from below the surface can be regarded as highly energy-consuming event [10]. The protruding ends of Z-pins are dragged along the fracture surface and then lead to fiber damage. The matrix debris surrounding the pins also suggests that pins are ploughed into the surrounding laminate substrate, in which pins are pushed into the matrix resin through shearing force.

Upon comparing the micrographs of pins with different diameters, it is found that thinner Z-pins tend to leave less fiber and resin residues in the holes after being pulled out, indicating that they experience less prominent "snubbing" friction compared to thicker pins. The higher length-diameter ratio of thinner Z-pins makes them are more flexible and are more likely to be completely pulled out. However, the reduction in Z-pin diameter also leads to a decrease in the load-bearing capacity of individual pin. This means that thinner pins are relative easier to fracture near the mid-plane due to the stress concentration caused by opening force. The sudden fracture of pins leads to the rapid release of accumulated energy, requiring high amount of energy. This phenomenon explains why the thinner pins exhibit higher fracture toughness compared to the thick ones.

## 2.2 Mode II Delamination Resistance

### *2.2.1 Materials Preparation and Experimental Setup*

T700 carbon fiber-epoxy prepreg supplied by Shanghai Cedar Co., Ltd is used to fabricate the specimens. The curved specimens are designed according to the geometrical model of a composite turbine blade with an inner arc length of 50.213 mm and a curvature radius of 157 mm. The dimension of the mode II test specimens is 150 mm × 25 mm × 4 mm, as illustrated in Fig. 2.15. The prepreg plies are laid up to a thickness of 4 mm with a stacking pattern of $[0/45/-45/90]_{4S}$. The prepreg stacks are debulked after every four plies to evacuate air between the plies. A PTFE film with length of 55 mm and thickness of 15 μm is imposed at the mid-plane of laminates during lay-up and a pre-crack is created for the crack initiation to ensure only bridging behavior of Z-pin is tested.

The PHZ process is applied to insert the pins into the uncured laminates. Firstly, a stainless needle is used to penetrate the uncured laminates with the support of a designated apparatus to ensure the verticality between the holes and the curved surface. The diameter of the needle varies depending upon the Z-pin diameter. The holes are uniformly distributed in the Z-pinned region. Then, the pins with the same length as the thickness of laminate are manually inserted into these holes. Finally, all the specimens are cured in an autoclave for 4 h at an overpressure of 0.6 MPa and a temperature of 120 °C. It is noticed that the prepreg is heated properly in advance to make Z-pins easy to insert within the softened resin matrix. Specimens are categorized into 9 groups. The specimens with thin pin (0.3 mm diameter) and thick pin (0.4 mm diameter) are marked with "a" and "A", respectively. The number followed by "a" and "A" represents the implant spacing of the pins. Detailed descriptions of the 9 groups of specimens are summarized in Table 2.1.

A three-point bending End Notched Flexure (ENF) test is used to measure the mode II interlaminar fracture of the unpinned and Z-pinned specimens. The experimental setup is illustrated in Fig. 2.15.

The tests are performed by an DNS100 universal machine with a calibrated 10 kN load cell. Prior to testing, a natural crack is propagated by 5 mm beyond the film insert edge by using a mode II load. A compression loading with a constant rate of 1.0 mm/min is applied to cause a delamination crack to grow along from the pre-crack. The crack propagation is recorded by a high-resolution camera capable

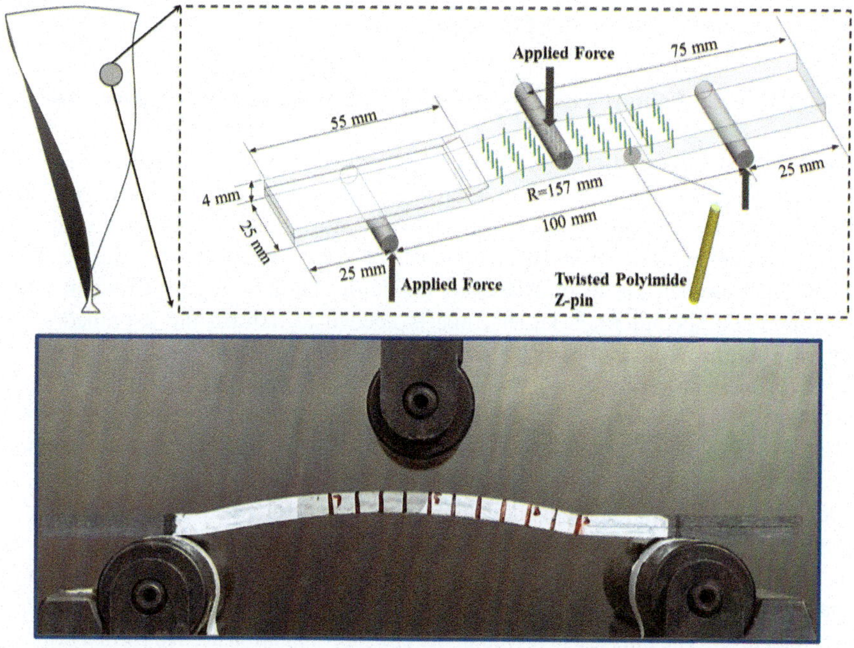


**Fig. 2.15** Schematic diagram and mode II loading process of curved laminates

**Table 2.1** Design parameters of specimens

| No. of specimen | Implant spacing (mm × mm) | Z-pin volume fraction (%) | Quantity |
|---|---|---|---|
| Unpinned | | | 3 |
| a3 | 3 × 3 | 0.7 | 3 |
| a4 | 4 × 4 | 0.4 | 3 |
| a5 | 5 × 5 | 0.3 | 3 |
| a6 | 6 × 6 | 0.2 | 3 |
| A3 | 3 × 3 | 1.3 | 3 |
| A4 | 4 × 4 | 0.7 | 3 |
| A5 | 5 × 5 | 0.5 | 3 |
| A6 | 6 × 6 | 0.3 | 3 |

of resolving the crack tip to a ± 0.5 mm accuracy. After the experiment, an electron digital microscope is used to observe the failure mode of specimens.

## *2.2.2 Experimental Results and Analysis*

### Effect of Z-pin Volume Fraction on Delamination Behavior of Laminates

Figure 2.16 shows the effect of Z-pin volume fraction on the mode II loading response. The load–displacement curves are intercepted when the crack propagation reaches 20 mm. As shown in Fig. 2.16, each curve starts with a liner growth, which indicates the specimens exhibit elastic deformation. For Z-pinned specimens, the deformation of the laminates is accompanied by the deformation of Z-pins. Compared with the unpinned specimens, a larger force is needed to deform the Z-pinned specimens. Therefore, the Z-pinned specimens exhibit an improved stiffness. However, the improvement produced by the presence of Z-pins would be partially compensated by the formation of resin-rich zone. Since the percentage of resin-rich zone is related to both the volume fraction and diameter of Z-pin, the stiffnesses of Z-pinned specimens are varied with different Z-pinning parameters. After a liner response, the specimens with Z-pin show a sudden load drop then followed by a period of fluctuation. However, this phenomenon does not exist in unpinned specimens. This can be attributed to the resin-rich zone introduced by the implantation of pins. When the crack propagates to the resin-rich zone, the weaker material properties of resin lead to a load drop. The pins are involved in the crack propagation area to form a bridging area. It can be observed that after the fluctuation, the load begins to rise steadily due to the formation of the bridging area. It is clearly shown that the existence of Z-pins improves the interlayer stiffness of specimens. In addition, it can be found that the interlayer stiffness improves with the increasing of Z-pin volume fraction.

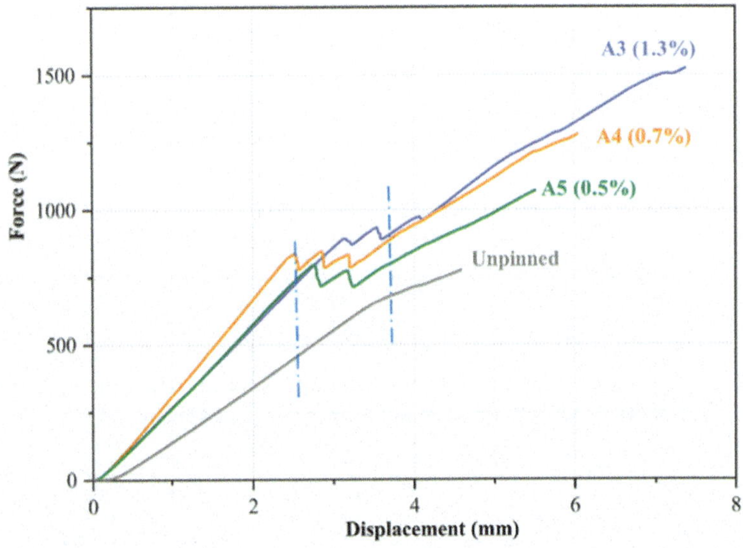


**Fig. 2.16** Load–displacement curves for specimens with different Z-pin volume fraction

The mode II fracture toughness can be represented by the critical strain energy release rate, which is calculated by Pegorin et al. [11]

$$G_{IIC} = \frac{9a^2 F \delta_{max}}{2B(2L^3 + 3a^3)} \tag{2.11}$$

where $a$ is the delamination length; $F$ is the applied bending force; $\delta_{max}$ is the deflection caused by $F$; $B$ is the width of the specimen; and $L$ is the support span length.

The calculated fracture toughness of different specimens is presented in Fig. 2.17. It can be found that the fracture toughness is significantly improved by Z-pinning. Compared with the unpinned specimen, the fracture toughness of the Z-pinned specimen is improved by up to 200%.

Furthermore, the relationship between Z-pin volume fraction and fracture toughness is shown in Fig. 2.18. With the same Z-pin diameter, the fracture toughness of the specimens increases linearly with Z-pin volume fraction.

The average velocity of crack propagation of each specimen is recorded as well, as shown in Fig. 2.19. The average velocity of crack propagation is calculated by $v = s/t$, where $s$ is the length of crack and $t$ is the time. It can be found that the crack propagation in the unpinned specimen develops rapidly, while in the Z-pinned specimen, the velocity of crack propagation is obviously reduced. The results indicate that the pin effectively slows down the crack propagation, and the velocity of crack propagation is reduced with the increase of Z-pin volume fraction.

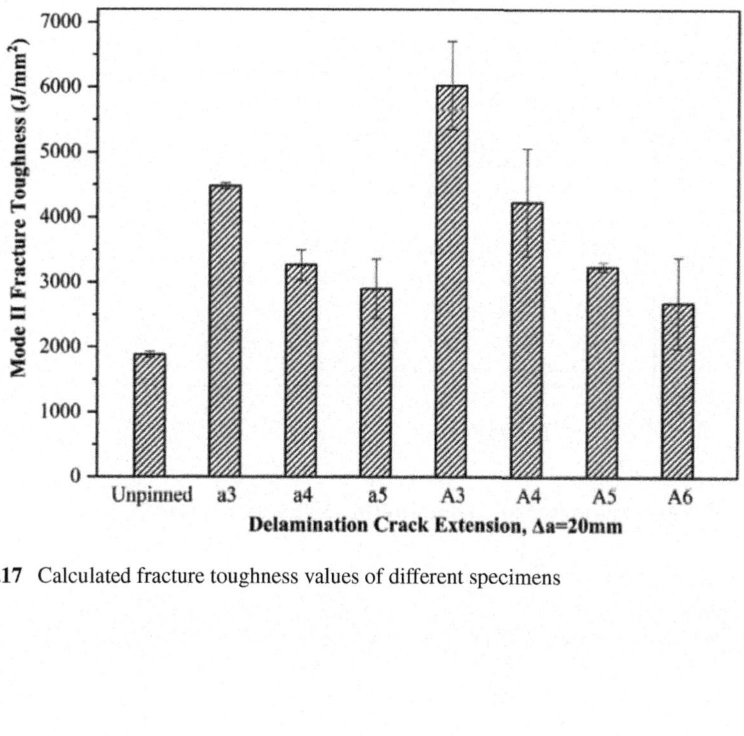


**Fig. 2.17** Calculated fracture toughness values of different specimens

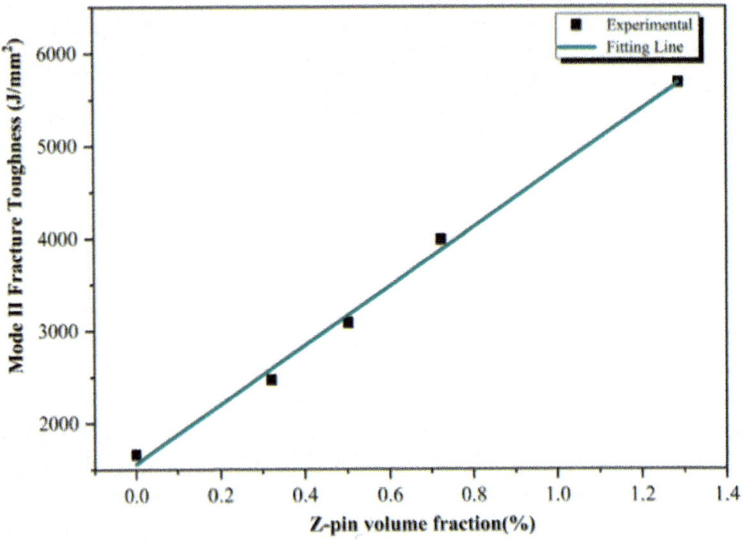


**Fig. 2.18** Relationship between Z-pin volume fraction and fracture toughness

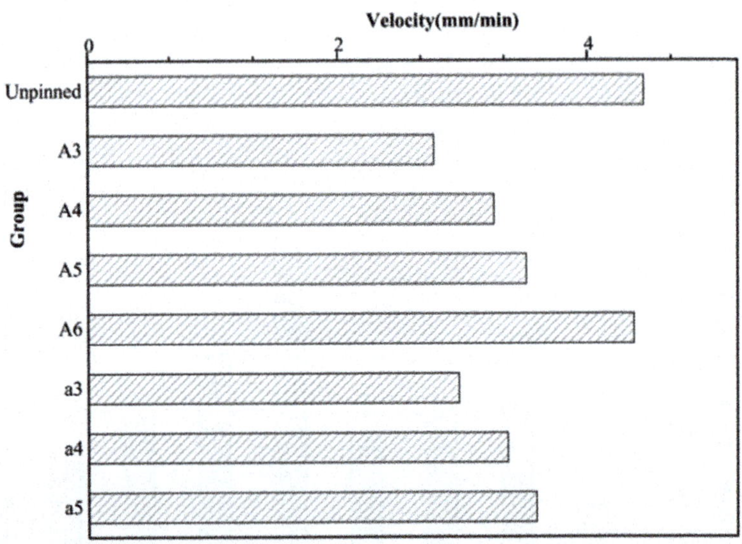


**Fig. 2.19** Velocity of crack propagation

## Effect of Z-pin Diameter on Delamination Behavior of Laminates

In order to explore the influence of Z-pin diameter on the delamination behavior of laminates, two sets of specimens of a3 and A4 with a volume fraction of 0.7% and a5 and A6 with a volume fraction of 0.3% are selected for comparison.

As shown in Fig. 2.20, the load–displacement curves are intercepted when crack is propagated by 20 mm. In the elastic deformation stage, the specimens with thick pin have larger interlayer stiffness. In the fluctuation stage the thick pin brings more load drop, which can be attributed to a larger resin-rich area caused by the thick pin.

The fracture toughness and velocity of crack propagation of specimens with different Z-pin diameter are listed in Table 2.2. With the same Z-pin volume fraction, the fracture toughness of specimens with thin pins is always higher than that of specimens with thick pins. Moreover, the velocity of crack propagation of Z-pinned specimens is reduced with the decrease of Z-pin diameter. The above data illustrates that under the same Z-pin volume fraction, the delamination resistance of specimens with thin pin performs better than those with the thick pin.

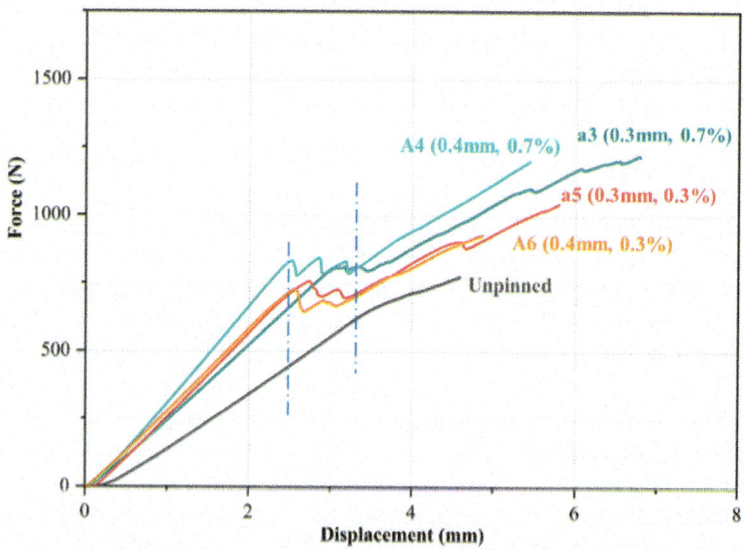


**Fig. 2.20** Load–displacement curves for specimens with different Z-pin diameter

**Table 2.2** Fracture toughness and crack propagation velocity of specimens with different Z-pin diameter

| No. of specimens | Z-pin diameter (mm) | Z-pin volume fraction (%) | Fracture toughness (J/m$^2$) | Velocity of crack propagation (mm/min) |
|---|---|---|---|---|
| Unpinned | | 0 | 1875.2 | 4.45 |
| a5 | 0.3 | 0.3 | 2575.5 | 3.59 |
| A6 | 0.4 | 0.3 | 2547.6 | 4.37 |
| a3 | 0.3 | 0.7 | 4446.5 | 2.97 |
| A4 | 0.4 | 0.7 | 4328.8 | 3.25 |

### Bridging Behavior and Failure Mode of Z-pin

During the mode II loading test, it is observed that the pins are either pulled out from the laminates or damaged. For unpinned specimens, the dissipated energy is entirely consumed by the surface energy of the newly created crack under mode II loading. For the Z-pinned specimens, in addition to the crack surface energy, the additional energy is required to pull-out or rupture the pins.

Figure 2.21 shows the morphologies of pins at different locations of the bridging zone. At the front of crack tip, the sliding distance between opposing crack face is small and pin is in elastic response without damage. With the crack propagation, pin begins debonding from the surrounding matrix. Then, pins are pulled out and the fibers of pin are split or even cut off by transverse shear force.

In mode II loading process, pins are mainly dominated by shear load and the failure process of both pins and laminate is complex, which is a combination of Z-pin pull-out, fiber fracture, resin fracture and resin crushing. The failure modes of pins mainly depend upon the sliding displacement of crack and the material of pin [12].

The shear failure types of Z-pins are captured in Fig. 2.22. During pull-out under mode II loading, pins can be crushed into the composite and then generate high friction traction stress, which induce traction load along bridging area and resist crack propagation. And the pair of shear forces are located on parallel axes and spaced a certain distance from each other. As the distance increases, the ratio of bending force to shear force increases. Besides, orthotropic property of pin makes it prone to failure in longitudinal splitting and pins are split into individual fiber bundles or individual fibers. As the pins go loose, the ratio of bending to shear loading increases further. Compared with the UAZ implantation method, in which the pin with chamfer head have tendency to be pulled out, the two ends of pins inserted by PHZ are symmetrical and pull out randomly.

In addition, fiber bridging inside pin and between pin and laminate is observed. During the curing process, the nesting of the adjacent layers fibers and weak interfaces induce the fiber bridging [13]. The presence of fiber bridging in the wake of a

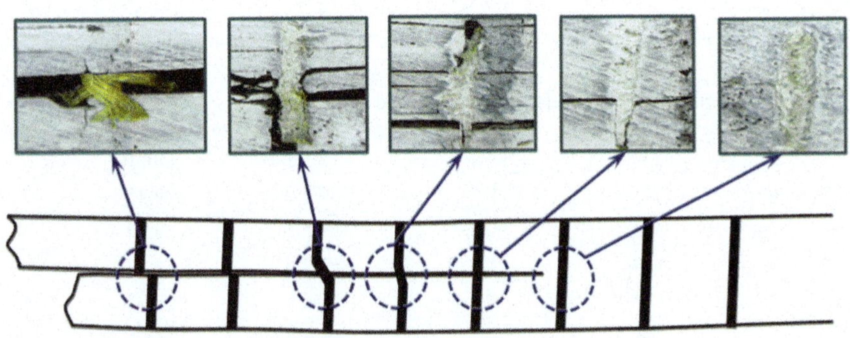

**Fig. 2.21** Morphologies of pins at different locations of bridging zone

(a) Internal shear failure; (b) Pull out; (c) Transverse shear failure; (d) Combination of transverse and internal shear failure.

**Fig. 2.22** Different failure modes of Z-pins under shear loading. **a** Internal shear failure; **b** pull out; **c** transverse shear failure; **d** combination of transverse and internal shear failure

delamination front can bridge crack surfaces and contribute to closure stress acting on these surfaces. Noteworthy, this phenomenon can be extended to Z-pinned laminate that beside the fiber bridging created by adjacent layers, the existing of pins generate the bridging behavior between fiber within Z-pin and between Z-pin and surrounding laminates (see Fig. 2.23). The fiber bridging behavior becomes resistance of crack propagation, which is beneficial to improving fracture toughness and slowing down the crack growth rate.

It is noticeable that during the loading process, when the crack reaches the tip of pre-crack of Z-pinned specimens, the delamination extends to the adjacent plies. This is caused by the void brought from the PTFE film [14]. At the same time, the change of interface configuration has an effect on the fiber bridging behavior. The specimens are quasi-isotropic, so the direction of crack propagation is not parallel to the fiber, thus the delamination deviates from the main crack plane into the neighboring plies to ensure the fiber bridging in the delamination plane. Compared with the blank specimen, the multiple cracks occur more frequently in Z-pinned specimens (see Fig. 2.24). And the insertion of Z-pin introduces more energy which may motivate the multiple cracks to dissipate the abundant energy [11].

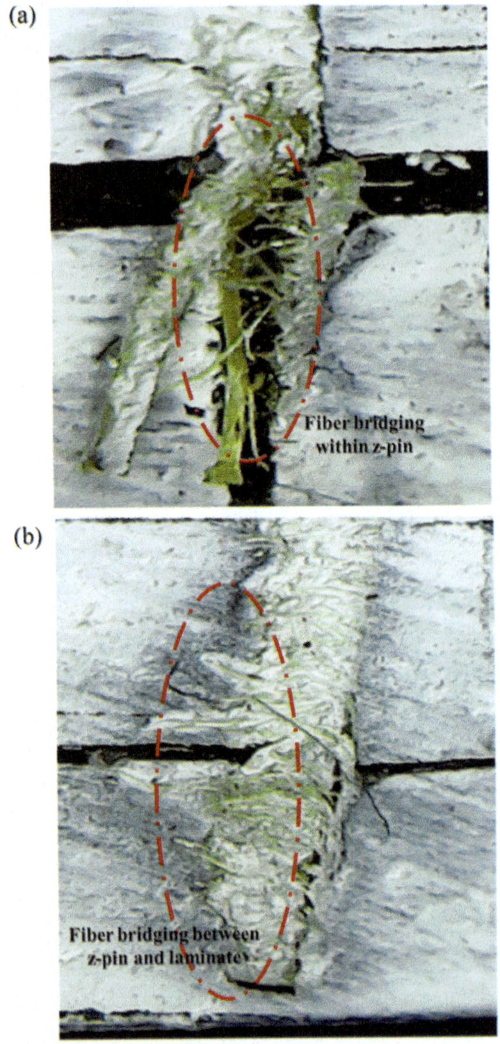


**Fig. 2.23** Fiber bridging behavior: **a** fiber bridging within Z-pin; **b** fiber bridging between Z-pin and laminate

## 2.3 Low-Velocity Impact and Post-Impact Properties

### *2.3.1 Experimental Setup*

The low-velocity impact experiments are conducted on INSTRON-9350 drop-weight impact tester in accordance with the ASTM D7136/7136M standard. T700 carbon fiber-epoxy prepreg supplied by Shanghai Cedar Co., Ltd is used to fabricate the specimens. The curved specimens are designed with an inner arc length of 50.213 mm and a curvature radius of 157 mm. The dimension of the specimens is 150 mm ×

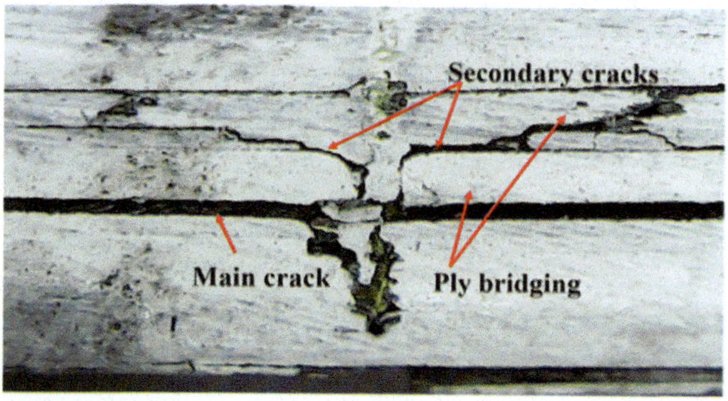


**Fig. 2.24** Multiple cracks and ply bridging

100 mm × 4 mm as Fig. 2.25 illustrated. The prepreg plies are laid up to a thickness of 4 mm with a stacking pattern of $[0/45/-45/90]_{4S}$. The PHZ process is applied to insert the pins into the uncured laminates. All the specimens are cured in an autoclave for 4 h at an overpressure of 600 kPa and a temperature of 120 °C. For simplicity, classified by the diameters and the spacing between Z-pins, the specimens are labeled as a/A-3/4/5/6, where a and A represent the thin (0.3 mm) and thick (0.4 mm) Z-pin, respectively, and the distance (mm) between the adjacent Z-pins is 3/4/5/6. For instance, a-3 indicates that the spacing between the thin Z-pins is 3 mm. Three specimens are fabricated in each group to ensure the consistence of the experiments.

The specimens are symmetrically clamped on a fixed base with a circular opening of a diameter of 75 mm. The impactor is hemispheric with a radius of 10 mm and the weight is 3.837 kg. The impact energy is set to 18 J. An anti-bouncing device is triggered after impact to prevent multiple impacts. After the low-velocity impact test, the damage on the specimen surface is recorded and the pit depths are measured using a micrometer. Then, a biaxial automatic ultrasonic C-scanner is used to measure the delamination damage projection area (DDPA) of the post-impact specimens.

The flexural after impact (FAI) tests are performed using DNS-100 universal machine according to ASTM D790 standard with a calibrated 10 kN load cell at a load rate of 2 mm/min. The FAI test specimens are cut to 150 mm × 25 mm × 4 mm to ensure the impacted region is in the center of the specimen. The support span is set to 100 mm, and the loading block is applied on the center of the specimen, as shown in Fig. 2.26. Once the contact force drops by more than 20%, the specimen is considered to fail and then the loading process is stopped. The bending process is recorded by a high-resolution camera.

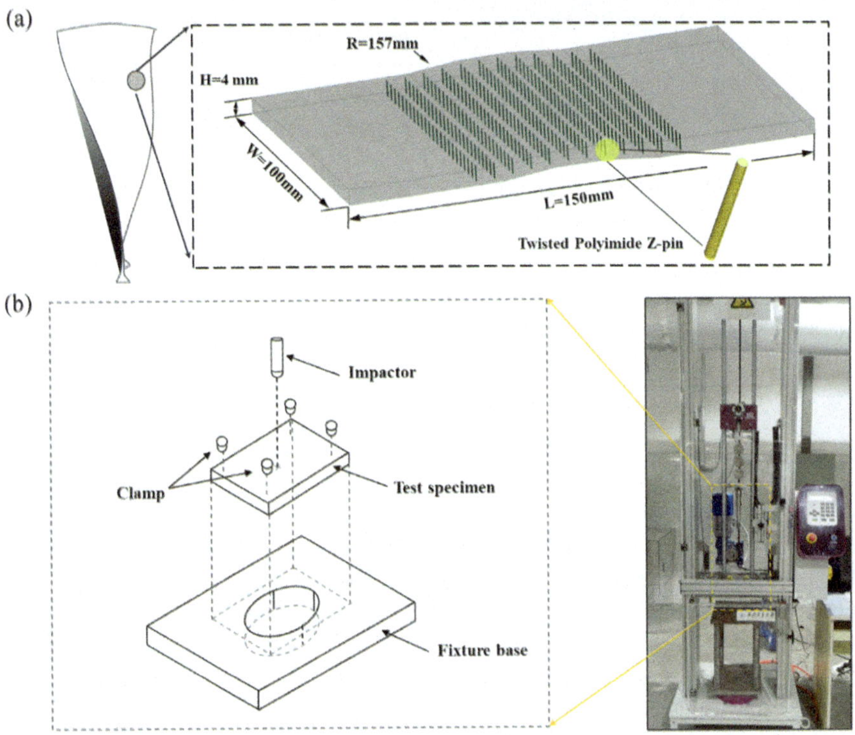


**Fig. 2.25** Schematic diagram of test equipment: **a** impact specimen; **b** impact setup

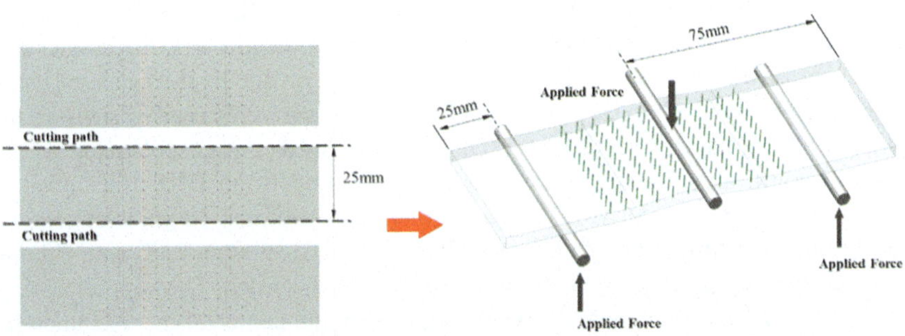


**Fig. 2.26** Schematic diagram of FAI specimens and test procedure

### 2.3.2 *Experimental Results and Analysis*

#### Impact Damage Modes and Characteristics of Low-Velocity Impact Response

The typical damage modes on the front and back surface of specimens with and without Z-pin are presented in Fig. 2.27. For unpinned specimens, dent is the mainly damage type on the impact side. And the matrix cracking induced by tensile loading occurs on the back side of the specimen. For Z-pinned specimens, besides dent on the impact surface, it is observed matrix crack in surface originates at the boundary of the impact region and the fiber fracture develops following paths perpendicular to the fiber direction. Z-pin appears to act as initiation site of fiber fracture in the impact side, due to the disruption generated in the laminate by the sharp needle during Z-pinning process, i.e., fiber breakage and fiber waviness. Moreover, Z-pins induces tensile cracking and multi-cracks along with Z-pins array on the back side. Additionally, pins lead to stress concentration as a foreign body and in-plane damage of the laminates, and then promote the onset of matrix cracks in the back side. The external damage of the specimens after impact, including dent depth and DDPA are summarized in Table 2.3.

The average dent depth is measured as residual depth of the pit after the impact force is removed. The residual dent depths of Z-pinned and unpinned specimens are 0.159 mm and 0.09 mm, respectively. The former is 76.7% larger than the latter. Z-pins are inserted into the laminates as a kind of foreign matter that destroys the integrity of the laminates and cause the fiber breakage, fiber waviness and resin-rich zone. So, these damages reduced the rebound of Z-pinned specimens compared to unpinned specimens. Besides, the dent depth of specimens with thin pin is slightly larger than that of the specimens with thick pins.

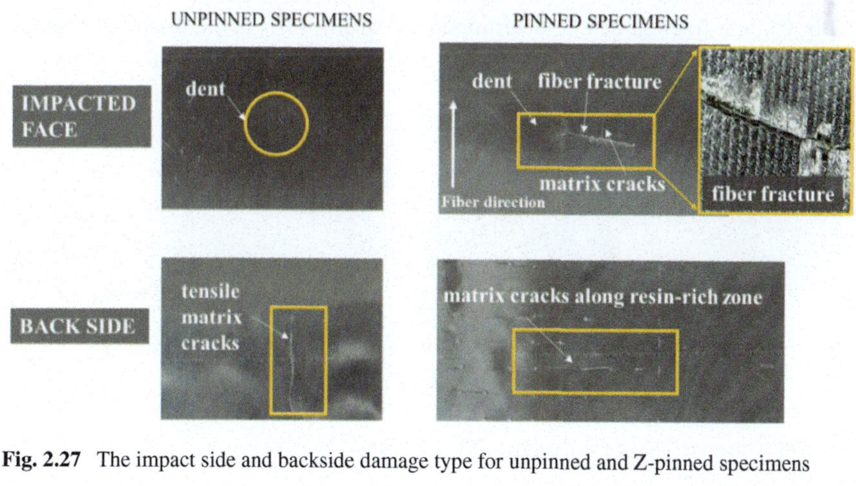


**Fig. 2.27** The impact side and backside damage type for unpinned and Z-pinned specimens

**Table 2.3** The impact damage results of the specimen

| No. of specimens | Z-pin spacing (mm) | Z-pin volume fraction (%) | Dent depth (mm) | DDPA (mm$^2$) |
|---|---|---|---|---|
| Unpinned | / | / | 0.09 | 1381.7 |
| a3 | 3 × 3 | 0.8 | 0.18 | 636.2 |
| a4 | 4 × 4 | 0.5 | 0.15 | 706.1 |
| a5 | 5 × 5 | 0.3 | 0.17 | 874.9 |
| a6 | 6 × 6 | 0.2 | 0.17 | 988.0 |
| A3 | 3 × 3 | 1.4 | 0.15 | 504.1 |
| A4 | 4 × 4 | 0.8 | 0.16 | 730.8 |
| A5 | 5 × 5 | 0.5 | 0.15 | 789.7 |
| A6 | 6 × 6 | 0.3 | 0.14 | 961.3 |

The size of delamination area is evaluated by calculating the number of pixels, as shown in Fig. 2.28. The scanning results give a colored map of cumulative damage as projected onto a horizontal plane. The green color represents damage exceeding 70% in the thickness direction. The yellow color represents the damage within 40–70%. The DDPAs of Z-pinned specimens are significantly lower than those of unpinned specimens. With 1.4% volume fraction of Z-pin, a reduction of DDPA by approximately 64% is achieved compared with the unpinned specimens. For Z-pinned specimens, once the delamination initiates, pins are activated to form bridging zones across the delamination cracks, which significantly slow down the developing of delamination and then reduces the damage area. During the impact, part of the impact energy is dissipated by energy-absorbing mechanisms of pins, i.e., Z-pin pull-out, pin/matrix debonding and Z-pin breakage. Furthermore, thin pins offer better delamination suppression performance than thick pins at the same Z-pin volume fraction. There are two variables in the presented tests: Z-pin diameter and Z-pin volume fraction. Z-pin volume fraction can be determined by $S_{\text{pin}}/S_{\text{la}}$ ($S_{\text{pin}}$ is the total cross-section area of pins and $S_{\text{la}}$ is implantation area). The $S_{\text{la}}$ is constant for all specimens, thus, to obtain the same volume fraction of Z-pin for the specimens with different pin diameters, the spacing must change to adjust the number of pins.

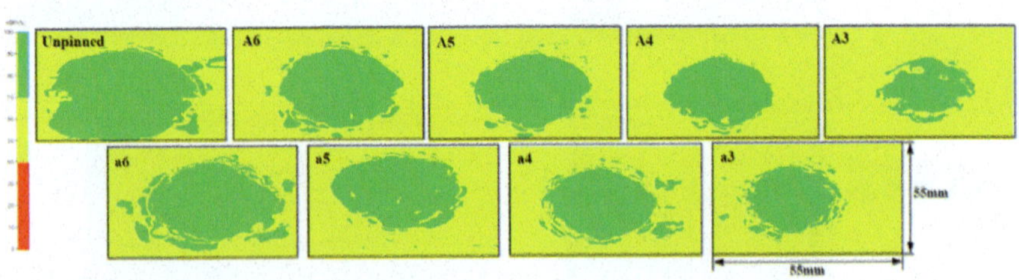


**Fig. 2.28** C-scan results of the specimens with different Z-pin parameters

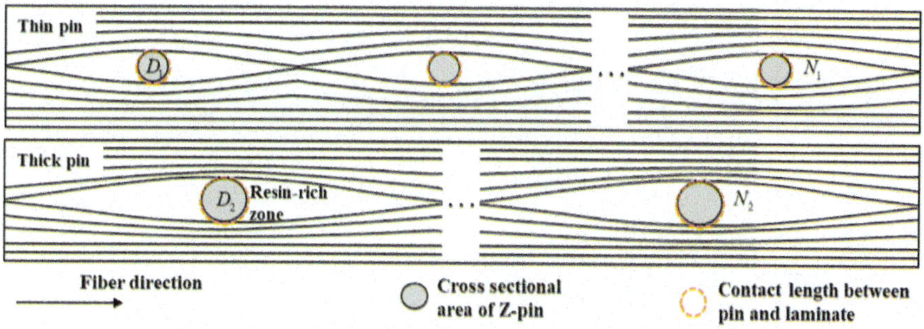


**Fig. 2.29** Schematic diagram of contact area between Z-pin and laminate

As illustrated in Fig. 2.29, to determine the bridging traction generated by the pins, the following equation [14] can be used:

$$T_b = N \int_0^l \tau(l)\pi D \, \mathrm{d}l \tag{2.12}$$

where $N$ is the number of pins, $\tau(l)$ is the interfacial shear strength of pin, $D$ is the pin diameter, and $l$ is Z-pin length. In this study, the length and interfacial shear strength of pins are constants for all specimens. Therefore, the bridging traction only depends on the number of pins and pin diameter.

Assuming the subscripts 1 and 2 represent the specimens with thick and thin pins with the same Z-pin volume fraction respectively. The following expression can be obtained:

$$N_1\pi D_1^2/4 = N_2\pi D_2^2/4 \tag{2.13}$$

The development of the above equation generates:

$$N_1 D_1 = N_2 D_2 D_2/D_1 \tag{2.14}$$

From the above equation, it is indicated $N_1 D_1 < N_2 D_2$. Thus, it can be concluded that larger bridging traction can be achieved by decreasing the pin diameter.

The low-velocity impact curves of an unpinned specimen are shown in Fig. 2.30. In the force–time curve, $F_{cr}$ is identified as critical threshold force that can be used to evaluate the initiation delamination resistance of laminates. At $F_{cr}$, the initial delamination of the composite occurs, and is thereafter accompanied by a rapid reduction in local flexural stiffness of the laminates. Thereby the load drops suddenly to a residual value $F_{re}$. Subsequently, a reloading to the overall peak force $F_{peak}$ happens for both unpinned and Z-pinned specimens as a sufficient residual potential energy is stored in the impactor [15]. Notably, the reloading phase is up to the overall maximum force $F_{peak}$ accompanied by high frequency noise. The high frequency

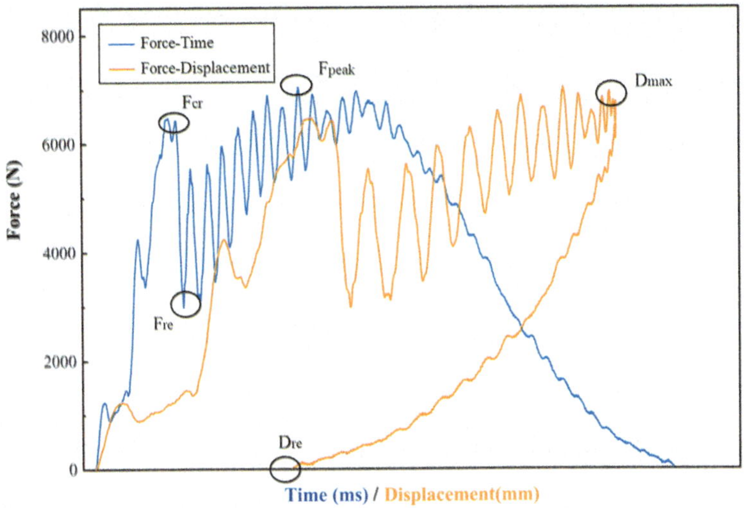


**Fig. 2.30** The low-velocity impact curves of an unpinned specimen

noise can be attributed to the release of energy caused by significant fiber fracture [16]. The specimen rebounding process, the load decreases to zero. In the displacement-force curve shows that the specimen continues to vibrate elastically after the impact but does not return to its initial position. The maximum displacement and residual deformation are marked as $D_{\max}$ and $D_{re}$, respectively.

## The Influence of Z-pin Volume Fraction on the Impact Resistance

The impact force–time curves of unpinned and pinned (with thick Z-pin) specimens are presented in Fig. 2.31. All the curves are similar and present high oscillations.

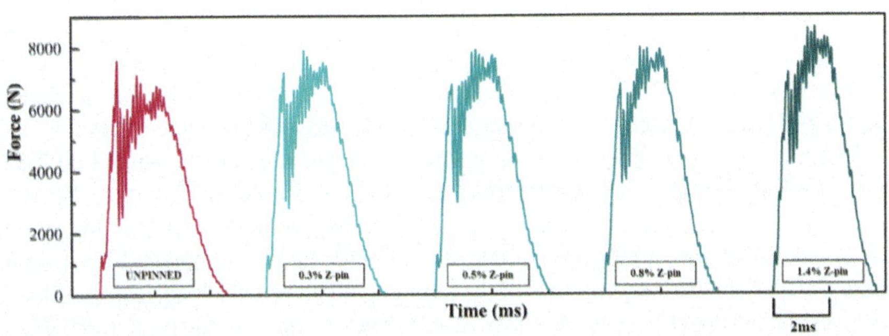


**Fig. 2.31** Force–time curves of unpinned and pinned specimens with different Z-pin volume fractions

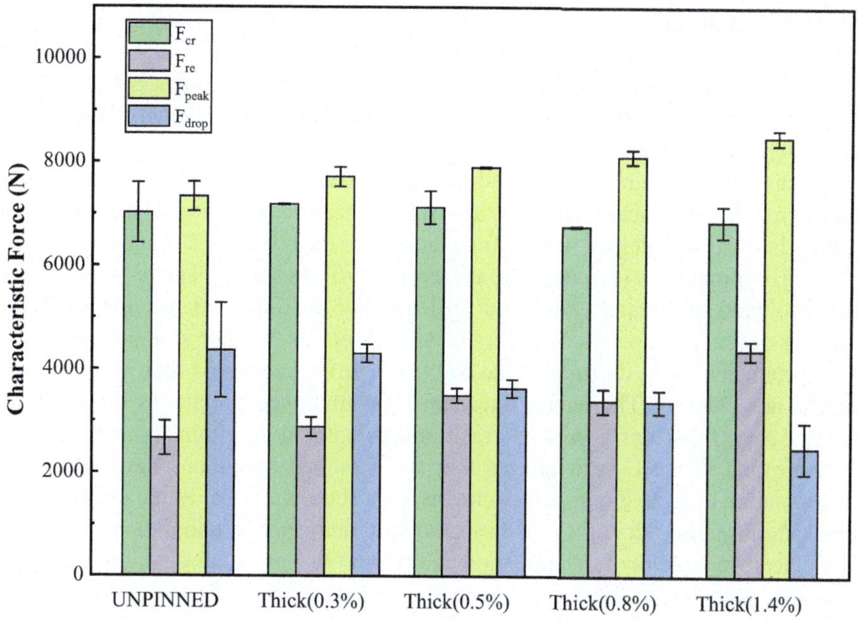


**Fig. 2.32** $F_{cr}$, $F_{re}$, $F_{peak}$ and $F_{drop}$ of unpinned specimens and pinned specimens with different Z-pin volume fractions

Initially, the effect of Z-pin is not obvious. During the delamination evolution, Z-pin gradually shows the toughening effect. $F_{cr}$, $F_{re}$, $F_{peak}$, and $F_{drop}$, where $F_{drop}$ denotes the difference between $F_{cr}$ and $F_{re}$, are plotted in Fig. 2.32. It can be found that both the residual force $F_{re}$ and maximum force $F_{peak}$ increase with the increase of Z-pin volume fraction, while the critical force $F_{cr}$ and the $F_{drop}$ show the opposite trend.

It is worth mentioning that $F_{peak}$ concerns the specimen stiffness after the initial impact damage at $F_{cr}$. It is substantiated by the fact that there is a larger delamination damage area in unpinned specimens compared with the Z-pinned specimens, while Z-pinned specimens have better toughness to recover faster from impact loading compared with unpinned specimens. Thus, the Z-pinned specimens can reach a high peak force after damage initiation. Besides, the small value of $F_{drop}$ indicates that the existing of Z-pin makes sure that the laminates suffer less damage and remain better performance under impact loading. However, the value of $F_{cr}$ for Z-pinned specimens is lower than the unpinned specimens. This phenomenon confirms that Z-pin leads to a reduction in the in-plane properties due to inevitable micro-damage and cannot improve the resistance to delamination initiation [17].

It is also notable that unpinned specimens have longer contact duration than Z-pinned specimens, being more obvious with increasing Z-pin volume fraction. This is because the laminates are damaged during impact, and the damage reduces the transverse combined stiffness of laminates which directly determines the acceleration of the impactor during impact. The longer contact duration implies a reduction in ultimate flexural stiffness due to more internal delamination damage. The toughening effect of Z-pin bridges the delamination cracks and then reduces stiffness degradation.

The impact force–displacement curves of unpinned and Z-pinned (thick Z-pin) specimens with different Z-pin volume fraction are presented in Fig. 2.33. In the force–displacement curves, unpinned specimens have the lowest maximum force and largest residual displacement, which are ascribed to the low stiffness of the specimens and endure the impact load in a ductile manner. On the contrast, Z-pinned specimens show higher maximum force and smaller residual displacement. The responses mean that Z-pinned specimens have higher stiffness. For the unpinned specimens, ductile manner under the impact load makes bending is easier to occur along the length of the specimens, more damage in the form of matrix cracks and then delamination cracks are induced. The energy mechanism in unpinned specimens dissipate the impact energy by newly created crack and delamination within the laminates. It is confirmed in the C-scan images that the Z-pinned specimens have much less delamination area. In Z-pinned specimens, pins act in the wake of the delamination crack front, nailing the crack as they gradually deformation and debonding. Z-pin limit the development of delamination and convert the energy dissipation into resin-rich zone with low stiffness. The energy dissipated in Z-pinned specimens is mainly through less delamination, fiber fracture positioned at the impact surface and matric cracks along the resin-rich zone on the back face.

The energy absorption curves of unpinned and pinned specimens with different Z-pin volume fraction are given in Fig. 2.34. The absorbed energy refers to the total energy dissipated by the specimen through the damage evolution. At $t = 0$ ms, the impactor begins to contact the specimens, whereby the kinetic energy of the impactor is delivered to the specimens. Part of the energy is stored in specimens in the form of elastic bending deformation and the rest is mainly dissipated by damage evolution, with negligible amount of energy dissipated through friction and heat. When the

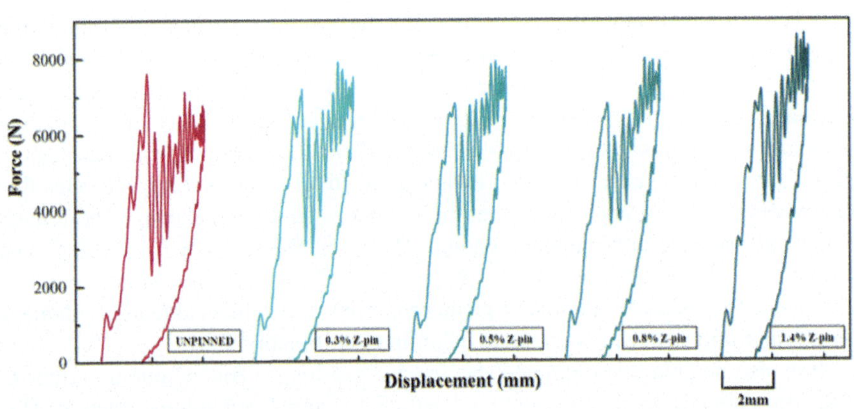


**Fig. 2.33** Force–displacement curves of unpinned specimens and pinned specimens with different Z-pin volume fractions

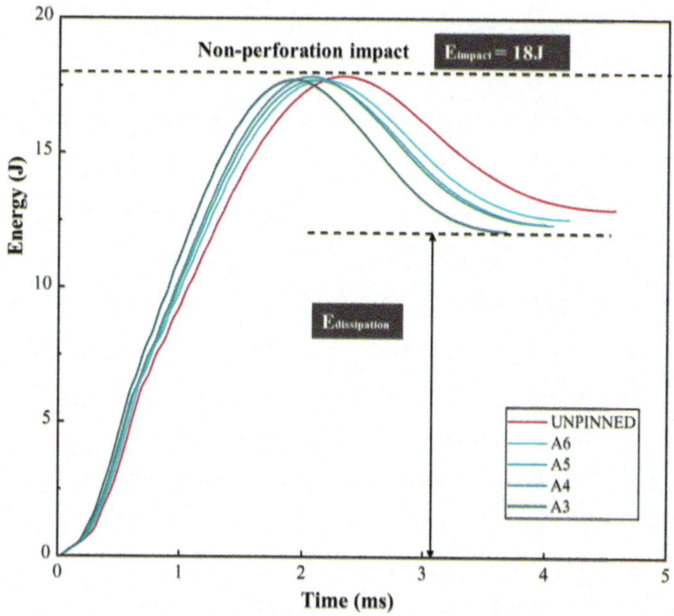


**Fig. 2.34** Transferred energy-time curves of unpinned specimens and pinned specimens with different Z-pin volume fractions

velocity of the impactor decreases to zero, a turning point of the energy absorption curves occurs, which means the impactor begins to rebound and the energy has been completely transferred to the specimens.

The energy dissipation rate $\Lambda$, which is expressed as $\Lambda = E_{dissipation}/E_{impact}$, where $E_{dissipation}$ is the total energy dissipation and $E_{impact}$ is the impact energy, is presented in Table 2.4. The results indicate that the value of $\Lambda$ decreases with the increasing of Z-pin volume fraction. Since the energy dissipation is equal to the energy absorbed through the damage of the specimens, Z-pin decreases the damage of laminates and resists damage evolution obviously.

The beneficial role played by Z-pinning in containing the spread of damage area is evident, whose delamination are extended to activate and exploit the closing action exerted by pins on delaminated interfaces. With the increasing of Z-pin volume fraction, a more continuous large-scale bridging zone is formed. Once a crack has grown small displacement into the Z-pin bridging zone, those pins will nail the delamination crack. So, the effectiveness of pins improves with Z-pin volume fraction. The mechanism of this bridging effect is the pin pullout process during crack propagate, which includes the elastic energy of Z-pins, the debonding between Z-pin and laminates and frictional work consumed by pulling out. Meanwhile, traction force to the crack surface is applied by pins. This process absorbs considerable amount of the stored strain energy that would otherwise be dissipated for delamination growth. The

Table 2.4 The energy dissipation rate for all the specimens

| No. of specimens | Z-pin volume fraction (%) | $E_{dissipation}$ (J) | $E_{dissipation}/E_{impact}$ (%) |
|---|---|---|---|
| Unpinned | / | 12.99 | 72.17 |
| a3 | 0.8 | 12.33 | 68.50 |
| a4 | 0.5 | 12.42 | 69.00 |
| a5 | 0.3 | 12.54 | 69.67 |
| a6 | 0.2 | 12.83 | 71.28 |
| A3 | 1.4 | 12.18 | 67.67 |
| A4 | 0.8 | 12.44 | 69.11 |
| A5 | 0.5 | 12.60 | 70.00 |
| A6 | 0.3 | 12.66 | 70.33 |

advantage is of the fact that the energy absorption capacity of delamination cracking of resin itself is considerably lower than that of Z-pin debonding and pullout.

#### The Influence of Z-pin Diameter on the Impact Resistance

Figure 2.35 presents impact force–time curves of unpinned specimens and Z-pinned specimens with different diameter. In general, each curve has a similar shape. It is shown that with the same Z-pin volume fraction, specimens with thin pin generate higher $F_{cr}$, $F_{re}$ and $F_{peak}$.

Additionally, $F_{drop}$ of specimens with thin pins is lower than that of specimens with thick pins (as illustrated in Fig. 2.36). The results reveal that Z-pin with smaller diameter perform better in resisting impact damage. This is due to the thin pin increases the contact area with the laminates and thin pin brings less micro-damage, i.e., Z-pin offset angle and the area of eyelet zone. The increase of interfacial contact

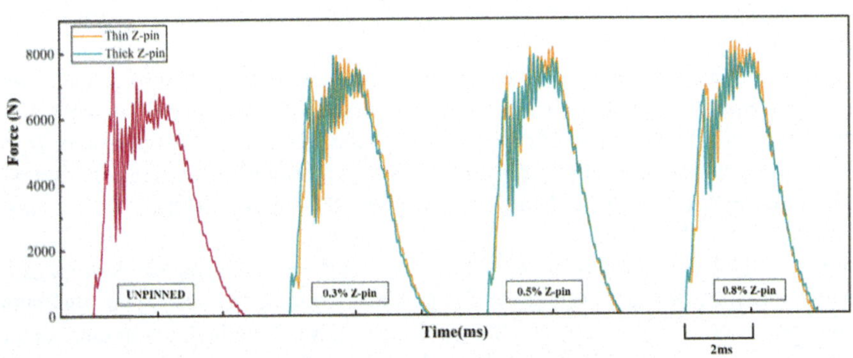


Fig. 2.35 Force–time curves of unpinned specimens and Z-pinned specimens with different Z-pin diameters

area reduces shear stress acting on each pin, which makes thin pin generates larger traction force than thick pin.

Figure 2.37 shows the impact force–displacement curves of unpinned specimens and Z-pinned specimens with different Z-pin diameter. In the stage of damage initiation and evolution, the load drop occurs earlier in the specimens with thick Z-pin due to more severe in-plane damage introduced by thick pin. It is noted that with the increasing of Z-pin volume fraction, the value of maximum displacement and residual displacement of specimens with thin and thick Z-pin are getting closer.

Figure 2.38 shows the energy-time curves of unpinned and Z-pinned specimens with different diameter. Under the same volume fraction, the values of Λ of the

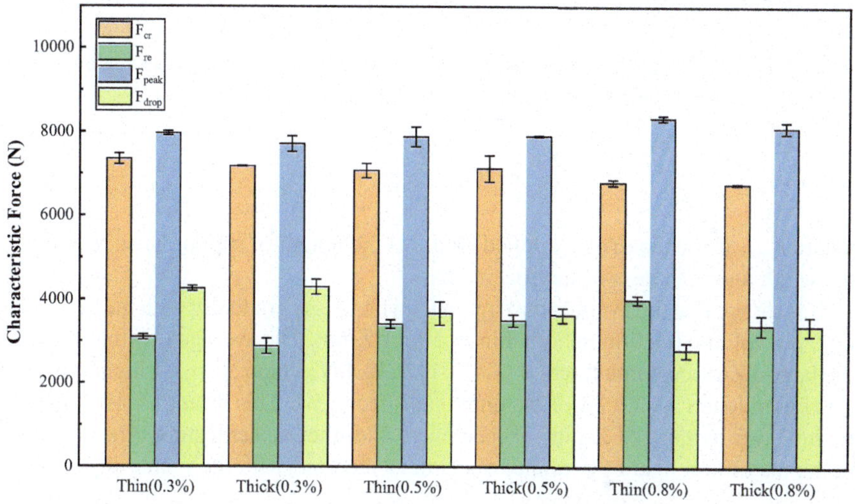


**Fig. 2.36** $F_{cr}$, $F_{re}$, $F_{peak}$ and $F_{drop}$ of Z-pinned specimens with different Z-pin diameters

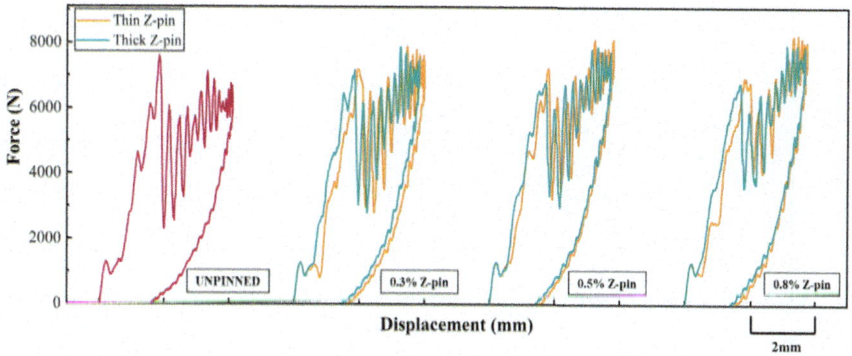


**Fig. 2.37** Force–displacement curves of unpinned specimens and Z-pinned specimens with different Z-pin diameters

specimens with thin Z-pin are lower than that of the specimens with thick pin. From the energy perspective, thin Z-pin can reduce the damage under impact loading, but the effect is slight.

As discussed above, the traction force provided by Z-pins is directly affected by the contact area between pins and laminate. Under the same volume fraction, Z-pin with thicker diameter has less contact area, which reduces the enhancing effect on delamination resistance. Additionally, thicker pins are arranged in a longer interval, thereby the large bridging zone is not as continuous as that of thinner pins.

All the specimens are bend to failure under the three-point bending load. During the FAI test, the specimens are mainly failed due to the delamination growth initiating at the impact damage area. The results of the maximum load $load_{max}$, deflection at $load_{max}$ and the flexural strength and modulus of different specimens are presented in Table 2.5.

The flexural strength, which refers to the critical stress on the outer surfaces layer of the specimen at failure, is determined by Chang et al. [18]:

$$\sigma = \frac{3F_{max}l}{2bt^2} \tag{2.15}$$

where $F_{max}$ is the maximum applied load; $l$ is the length of support span; $b$ and t are the width and thickness of the specimens.

It is shown that the average maximum load increases gradually with the increasing of Z-pin volume fraction in both thin and thick Z-pin. The average maximum load at the largest Z-pin volume fraction (1.4%) is 3632.4 N, which is 76.1% higher than the maximum load (2062.7 N) of the unpinned specimens. This dramatic rise attributes to the lower damage of Z-pinned specimens after the impact testing. The deflection at the maximum load of Z-pinned specimens increase with the increasing of Z-pin

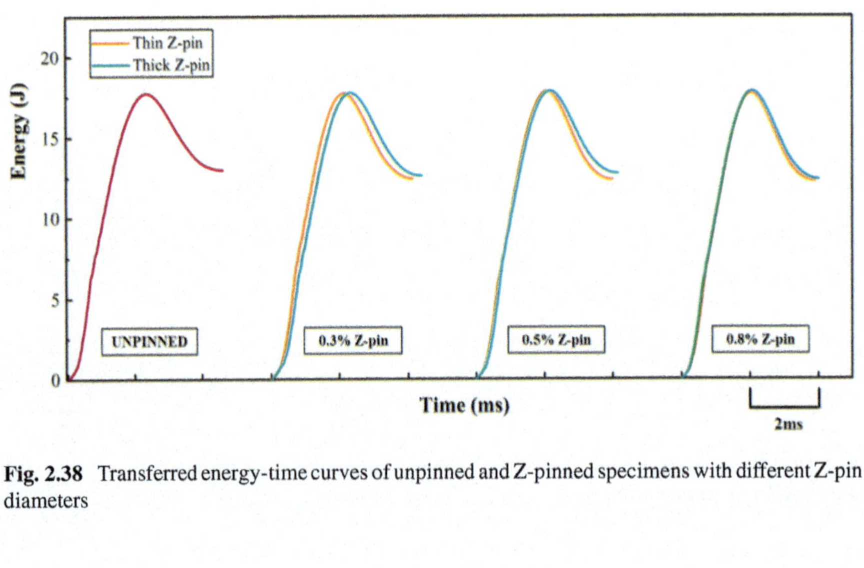


**Fig. 2.38** Transferred energy-time curves of unpinned and Z-pinned specimens with different Z-pin diameters

**Table 2.5** Comparison of the FAI properties of different specimens

| No. of specimens | Z-pin volume fraction (%) | $load_{max}$ (N) | Deflection at $load_{max}$ (mm) | Flexural strength (MPa) |
|---|---|---|---|---|
| Unpinned | / | 2062.7 | 4.59 | 464.11 |
| a3 | 0.8 | 2981.4 | 6.28 | 670.82 |
| a4 | 0.5 | 2879.8 | 6.46 | 647.96 |
| a5 | 0.3 | 2675.4 | 5.94 | 601.97 |
| a6 | 0.2 | 2436.3 | 4.69 | 548.17 |
| A3 | 1.4 | 3632.4 | 8.31 | 817.29 |
| A4 | 0.8 | 2958.5 | 7.43 | 665.66 |
| A5 | 0.5 | 2673.9 | 5.43 | 601.63 |
| A6 | 0.3 | 2613.7 | 5.8 | 588.08 |

volume fraction. For the specimens with the largest Z-pin volume fraction (1.4%), the deflection has an improvement of 80.88% compared to the unpinned specimens. In addition, it can be found that Z-pin has a significant effect on the improvement of the flexural strength.

It is worth mentioned that the existence of Z-pin not only increases the flexural strength, but also slows down the damage evolution. The stage of damage evolution is illustrated in Fig. 2.39, which is from the first load peak (the visible damage appears) to final failure of the specimen. The length of this stage for the specimens with the highest Z-pin volume fraction is about twenty times larger than that of unpinned specimens. It is shown that Z-pinned specimens do not fail immediately after reaching the limit load but existing a stage that the load decreases gradually. This phenomenon provides a guarantee for the safety of the composite structures in the actual applications.

The failure process of the unpinned and pinned specimens in bending tests are shown in Fig. 2.40. For unpinned specimens, the laminate failed by bending-induced tensile rupture of the surface ply, which was accompanied by delamination. At the first stage, the flexural force–displacement curves remained linear to ultimate failure, suggesting that the delamination on the tensile side was not critical. The destructive delamination did not form until after the main fracture event.

For Z-pinned specimens, the composite also failed by tensile rupture of the surface ply, but the delamination cracking was less extensive compared with unpinned specimens. Z-pins suppressed the formation of multiple delamination cracks by generating bridging tractions that reduce the stress at the crack tip. The suppression effect increases significantly with Z-pin volume fraction. It is observed that delamination cracking mainly occurred at interlayers near and below the mid-plane, which experienced tensile stress. It is surmised that the cracks were initiated in clusters of broken fibers near Z-pins, spread radially towards similar cracks propagating from neighboring pins, and coalesced into a through-width crack that caused final failure. The crack propagation process was accompanied by the activation of Z-pin columns in the order of their distance to the initial damage.

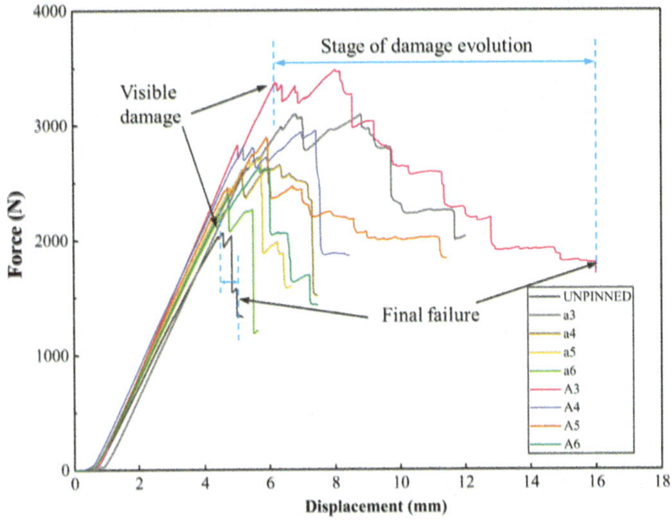


**Fig. 2.39** Force–displacement curves of flexural after impact

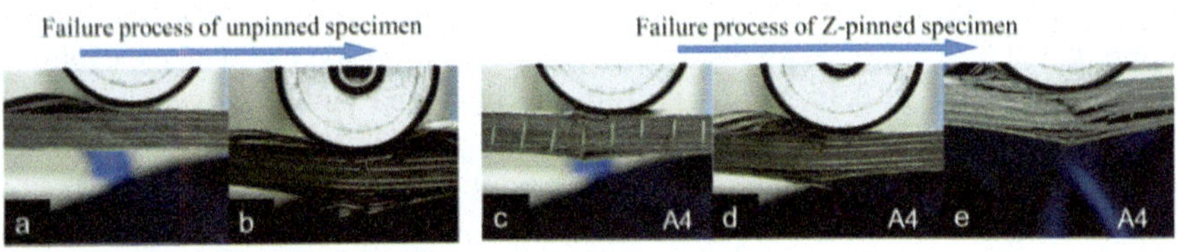


**Fig. 2.40** Failure process of unpinned and Z-pinned specimens

## 2.4 Conclusions

In this chapter, experimental tests are conducted to study the influence of Z-pining parameters on the mode I and mode II interlaminar delamination resistance properties, low-velocity impact and post-impact flexural properties on Z-pinned CFRP composites. The specimen design, experimental setup and test procedure are introduced in detail. The efficiency of Z-pining on interlaminar toughening effect is investigated based on the experimental results:

(1) For mode I delamination resistance, the insertion of Z-pins leads to an improvement of interlaminar fracture toughness due to the bridging mechanism. With a Z-pin volume fraction of 0.8 vol%, an improvement of $G_C$ value is 567% compared with unpinned cases. The delamination resistance capacity of Z-pins is dependent on the mixed-mode ratio, which is due to the transition in delamination resistance mechanism and failure behavior of the Z-pins.

(2) For mode II delamination resistance, the toughening effect of Z-pins is substantial. Z-pinning increases the interlayer stiffness of laminates. The mode II interlaminar delamination toughness improves with the increase of Z-pin volume fraction and has a linear relationship with Z-pin volume friction. Under the same volume fraction, pin with smaller diameter can prevent crack propagation more effectively by increasing the contact area between pins and laminates. The failure process of Z-pinned laminate includes a complex combination of lateral and longitudinal displacement, a superposition of crack growth resistance of the resin matrix and multiple failure micro-mechanisms of the Z-pin bridging behavior.

(3) Compared with the unpinned specimens, increasing Z-pin volume fraction can result in an improvement of impact resistance accompanied by larger residual dent depth, less matrix cracks and smaller DDPAs. The DDPAs of Z-pinned specimens are 28.5–63.5% smaller than unpinned specimens.

(4) Z-pining could improve the residual flexural strength significantly. The FAI test results show that the residual flexural strength of specimens with high Z-pin volume fraction is 18.1–76.1% higher than that of the unpinned specimens.

## References


1. Mouritz AP, Review of Z-pinned composites laminates, Composites Part A: Applied Science and Manufacturing. 2007;38: 2383–2397.
2. Mouritz AP, Review of Z-pinned CFRP composites and sandwich composite, Composites Part A: Applied Science and Manufacturing. 2020;139: 106128.
3. Yasaee M., Lander J.K., Allegri G. Hallett SR. Experimental characterisation of mixed mode traction–displacement relationships for a single carbon composite Z-pin, Composites Science and Technology 2014; 94: 123–131.
4. Yasaee M., Bigg L., Mohamed G., Hallett S.R., Influence of Z-pin embedded length on the interlaminar traction response of multi-directional composite laminates, Materials and Design. 2017;115: 26–36.
5. Guedes R.M., A. Sa, Moura M.F.S.F.d., An experimental and numerical assessment of DCB tests on glass/polyester curved beams cut out from pipes, Polymer Testing. 2008; 27: 985–994.
6. A. Ghadirdokht, M. Heidari-Rarani, Delamination R-curve behavior of curved composite laminates, Composites Part B: Engineering. 2019; 175: 107139.
7. S. Hashemi, A.J. Kinloch, J.G. Williams. The Effects of Geometry, Rate and Temperature on the Mode I, Mode II and Mixed-Mode I/II Interlaminar Fracture of Carbon-Fibre/Poly(ether-ether ketone) Composites, Journal of Composite Materials. 1990; 24: 918–956.
8. J.W. Hutchinson, H.M. Jensen, Models of fiber debonding and pullout in brittle composite with friction, Mechamcs of Materials. 1990; 9: 139–163.
9. S. Zhang, Y. Xu, W. Zhang, A novel micromechanical model to study the influence of cure process on the in-plane tensile properties of Z-pinned CFRP composites, Composite Structures. 2022; 300: 116156.
10. A.D. Cochrane, J. Serra, J.K. Lander, H. Böhm, T. Wollmann, A. Hornig, M. Gude, I.K. Partridge, S.R. Hallett, Experimental investigation ofhigh strain-rate, large-scale crack bridging behaviour of Z-pinned tapered laminates, Composites Part A: Applied Science and Manufacturing. 2022;155: 106825.

11. F. Pegorin, K. Pingkarawat, S. Daynes, A.P. Mouritz, Mode II interlaminar fatigue properties of Z-pinned carbon fibre reinforced epoxy composite, Composites Part A: Applied Science and Manufacturing. 2014; 67: 8–15.
12. D.D.R. Cartie´, M. Troulis, I.K. Partridge, Delamination of Z-pinned carbon fibre reinforced laminates, Composites Science and Technology. 2006; 66: 855–861.
13. R. Khan, Fiber bridging in composite laminates: A literature review, Composite Structures. 2019; 229: 111418.
14. T.M. Koh, S. Feih, A.P. Mouritz, Experimental determination of the structural properties and strengthening mechanisms of Z-pinned composites T-joints, Composite Structures. 2011; 93: 2222–2230.
15. D.D.R. Cartie, P.E. Irving, Effect of resin and fibre properties on impact and compression after impact performance of CFRP, Composites Part A: Applied Science and Manufacturing. 2002; 33: 483–493.
16. U.A. Khashaba, R. Othman, Low-velocity impact of woven CFRE composite under different temperature levels, International Journal of Impact Engineering. 2017; 108: 191–204.
17. X. Zhang, L. Hounslow, M. Grassi, Improvement of low-velocity impact and compression-after-impact performance by z-fibre pinning, Composites Science and Technology. 2006; 66: 2785–2794.
18. P. Chang, A.P. Mouritz, B.N. Cox, Flexural properties of Z-pinned CFRP composites, Composites Part A: Applied Science and Manufacturing. 2007; 38: 244–251.

# Chapter 3
# Curing Process Modeling of Z-pinned CFRP Composites

Curing process after Z-pinning insertion becomes one of the essential issues affecting the enhancement of Z-pins. When Z-pinned prepreg laminates undergo elevated temperature and pressure loads in autoclave, residual stresses are gradually generated, which are attributed to thermal expansion and chemical shrinkage mismatch between different phases in the microstructure. Stress concentration is mostly accumulated in the interface neighboring Z-pins and ultimately exceeded the strength of the interface, which result in debonding behavior around Z-pins. Thereby, it is worth noting that the effect of curing process could not only generate residual stresses but also produce interfacial cracking around Z-pins, which definitely alters the bridging structures of Z-pins. To validate these assumptions, scanning electron microscope (SEM) is unitized to investigate the microstructural morphologies around Z-pins, and several in-situ morphologies have given appropriate evidence to prove interfacial cracking existed indeed after the cure. Clearly, interfacial debonding around Z-pins is an inherent problem caused by residual stresses in the cure. Since bridging response of Z-pins is incorporated the aspects of friction and adhesion, partially debonding after the cure should not be negligible.

In this chapter, a novel micro-mechanical modeling method is proposed to analyze the evolution of cure-induced residual stresses and its effect on the interfacial cracking of Z-pins after the cure. A representative volume element (RVE) model of the Z-pinned CFRP composites is developed. A curing reaction kinetics model is developed to predict the curing reaction of resin. Cohesive zone model (CZM) [1] is used to describe the adhesive behavior of the interface between Z-pin and resin. The curing process simulation is conducted to predict the evolutions of degree of cure (DoC) and temperature together with residual stresses of each phase (base laminate, resin pocket, Z-pin and interface). Then, regarding the predicted residual stress as a predefined boundary, the failure mechanism of the Z-pin/resin interface is investigated by coupling the quadratic stress (QUADS) criterion [2, 3] and the Benzeggagh and Kenane (B-K) propagation criterion [4]. Based on damage analysis, the specific location and amount of interfacial cracking are determined. The numerical simulation



Y. Xu and W. Zhang, *Modeling of Z-pinned Carbon Fiber-Reinforced Polymer (CFRP) Composite*, https://doi.org/10.1007/978-981-96-3628-0_3

results are validated against the experimental examination to justify the predictive capability of this proposed approach. Moreover, the effect of curing process parameters on the residual stresses development and interfacial cracking progression is also discussed.

## 3.1 Geometrical Model Generation

### *3.1.1 Specimen Fabrication*

Experimental tests are conducted to investigate the microstructural morphologies of the Z-pinned CFRP composites after the curing process. The geometry of the specimens is designed as 75 mm × 25 mm × 4 mm. The composite laminates are fabricated using a unidirectional T300/5228 prepreg with a fiber volume content of 56%. The density of the prepreg is 1560 kg/m$^3$ and the thickness is 0.125 mm. Prepared by the hand lay-up technique, the specimens containing 32 piles of prepreg are stacked into a $[0/45/90/45]_{4s}$ pattern with a quasi-isotropic character. Vacuum bag is applied to separate air connection within the lay-up prepregs and to pre-compact them as well. Prior to the cure, the central part of the specimens covering a square of 25 mm × 25 mm is reinforced with 49 Z-pins to a volume content of 1.5%, i.e., seven columns of Z-pins in the x-direction and seven rows of Z-pins in the y-direction. Z-pins are made by pultruding T300 carbon fibrous rods impregnated with bismaleimide (BMI) resin with a diameter of 0.51 mm. In order to give wonderful control of the implantation of Z-pins, a pre-hole Z-pinning (PHZ) technique is used, which consists of two steps: (a) the uncured prepregs stacks are placed on a warm plate (27–33 °C) and punched vertically by a fine needle to create pre-holes; (b) pins are cut into 4 mm long which is equivalent to the thickness of the prepreg laminates, and gently reinforce into the perforated holes for less Z-pin misalignments. Detail of the PHZ technique is described in this chapter. Afterwards, the prepreg laminates reinforced with Z-pins are placed in an autoclave and consolidated under the predetermined curing profile. The curing profile is given in Fig. 3.1. Firstly, the curing temperature is increased from the room temperature to the holding temperature 180 °C with a heating rate of 3 °C/min. Secondly, the temperature is keeping at a constant value of 180 °C for 120 min. Thirdly, the temperature has a decline from the holding temperature 180 °C to the room temperature with a cooling rate of 3 °C/min. In the meantime, the sustained pressure is set to be 0.6 MPa.

To summarize, the process of fabricating the Z-pinned specimens is illustrated in Fig. 3.2.

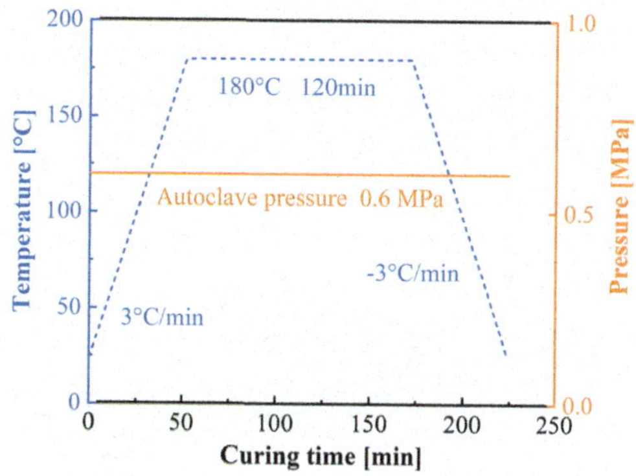


**Fig. 3.1** Curing temperature profile and pressure for Z-pinned specimens

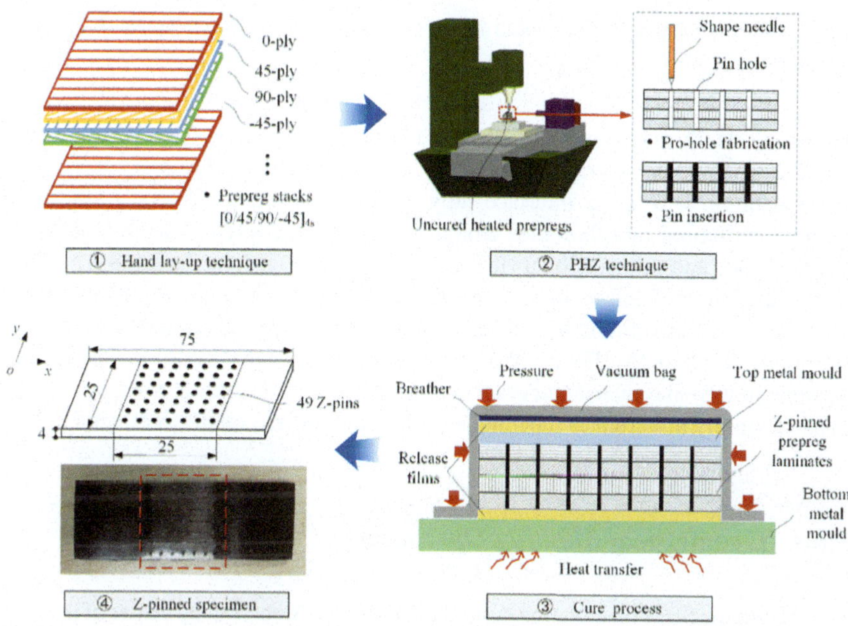


**Fig. 3.2** Flowchart of the fabricating of Z-pinned specimens

### *3.1.2 Experimental Characterization*

The cured Z-pinned specimens are examined through SEM approach. The in-plane view of the microstructural morphology around the Z-pin is presented in Fig. 3.3a. It can be observed that an eye-shape resin-rich zone is developed and elongated along the fiber direction of the base laminate. Fiber waviness is induced on both flanks of the resin-rich zone. This is owing that the fibers are spread laterally when they pass around the Z-pin, and thus the flowing resin can be full of the eye-shape gap. The dimensions of the local Z-pinned area are measured from photomicrographs of

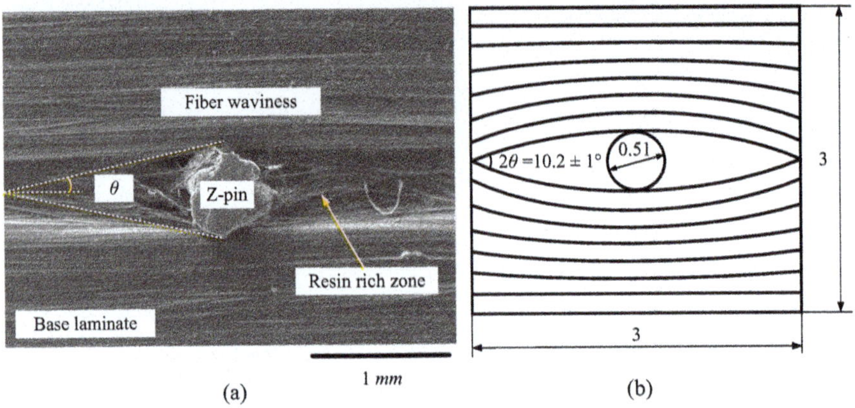


**Fig. 3.3** Illustrations of a single Z-pinned in composite laminates. **a** Photomicrograph. **b** Geometric description

specimens, as illustrated in Fig. 3.3b. The fiber distortion angle ($\theta$) is determined to be 5.1 ± 0.5° and the average length of the eye-shape zone is 3 mm.

To distinguish whether interface fails or matrix near interface fails after cure, the microstructural morphologies around Z-pins are investigated from both cross-sectional view and side-sectional view. As displayed in Fig. 3.4a, the Z-pin is partially debonded to the base laminate with interfacial cracking around it. Partially interfacial cracking can be also found by a side-sectional view, as presented in Fig. 3.4b from other open literature [5]. The SEM observations indicate that the cure-induced failure is dominated by the interfacial cracking and not matrix cracking. To this end, this research mainly investigates the influence of curing process on the interfacial failure.

### *3.1.3 Geometrical Shaping Operations*

Based upon the SEM observation, a single Z-pin is positioned at the center of the RVE and modeled as a circular part. The microstructures of fiber distortion and resin-rich pockets caused by Z-pins are explicitly assumed. Two reference values (i.e., local fiber distortion angle $\theta$ and fiber volume content $V_f$) are introduced here to detail the characteristics of the fiber distortion. Refer to Pierreux et al. [6], the maximum value of the fiber volume content can be determined as 0.90, which indicates a fully compacted fiber-reinforced zone created nearest the Z-pin. Resin-rich pocket is a pool of resin matrix that fills the voids induced by the fiber waviness during the cure process.

On the basis of microstructural assumptions, the mesoscopic geometrical characteristics of the RVE at an arbitrary cross-section are determined in Fig. 3.5a. A local coordinate axis system is positioned at the center of the Z-pin region. The $u$-axis is assumed to align with the fiber direction of the lamina, and the $v$-axis is vertical to the

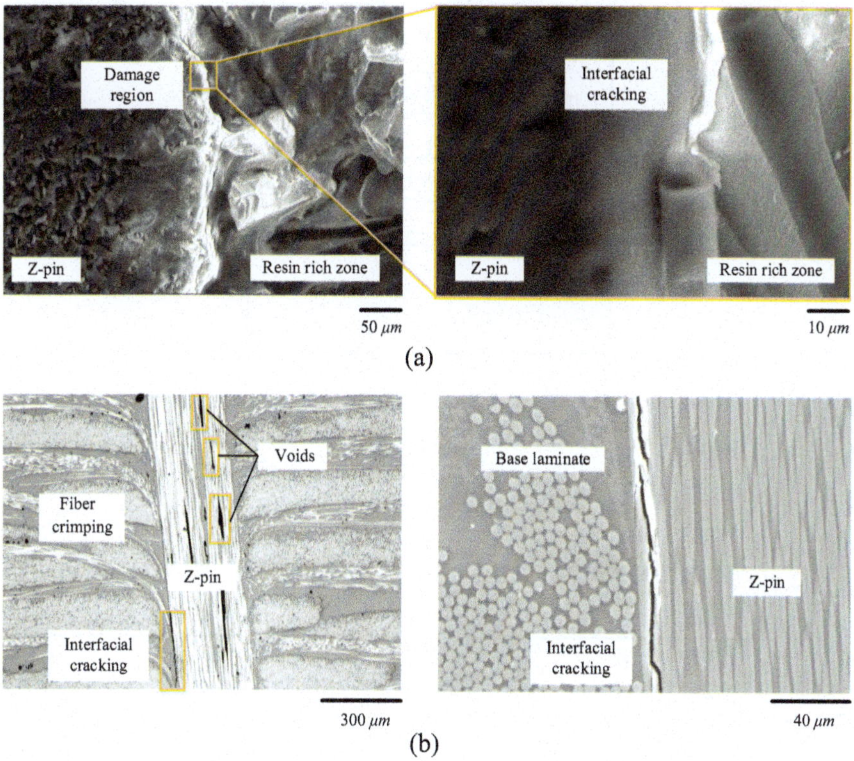


**Fig. 3.4** Partially debonding of a single Z-pinned in composite laminates. **a** In a cross-sectional view. **b** In a side-sectional view [5]

fiber direction. The geometry of the RVE model for inserting thin and thick Z-pins is respectively listed in Table 3.1. The fiber distortion angle $\theta(u, v)$ and the fiber volume content $V_f(u, v)$ at an arbitrary position of the cross-section are respectively assumed as follows:

$$\theta(u,v) = \begin{cases} \theta_{lam} & \frac{W}{2} < |v| \\ \theta_{lam} + \frac{2v-w}{w-W}\left(\theta_{max}(u) - \theta_{lam}\right) & \frac{w}{2} < |v| \le \frac{W}{2} \\ \theta_{max}(u) & |v| = \frac{w}{2} \end{cases} \tag{3.1}$$

$$V_f(u,v) = \begin{cases} V_{f,lam} & \frac{W}{2} < |v| \\ V_{f,lam} + \frac{2v-w}{w-W}\left(V_{f,bound}(u) - V_{f,lam}\right) & \frac{w}{2} < |v| \le \frac{W}{2} \\ V_{f,bound}(u) & |v| = \frac{w}{2} \end{cases} \tag{3.2}$$

where $\theta_{lam}$ and $\theta_{max}$ stand for the normal fiber distortion angle and maximum fiber distortion angle, respectively. $V_{f,lam}$ and $V_{f,bound}$ represent the fiber volume contents at the normal lamina and resin-rich boundary, respectively.

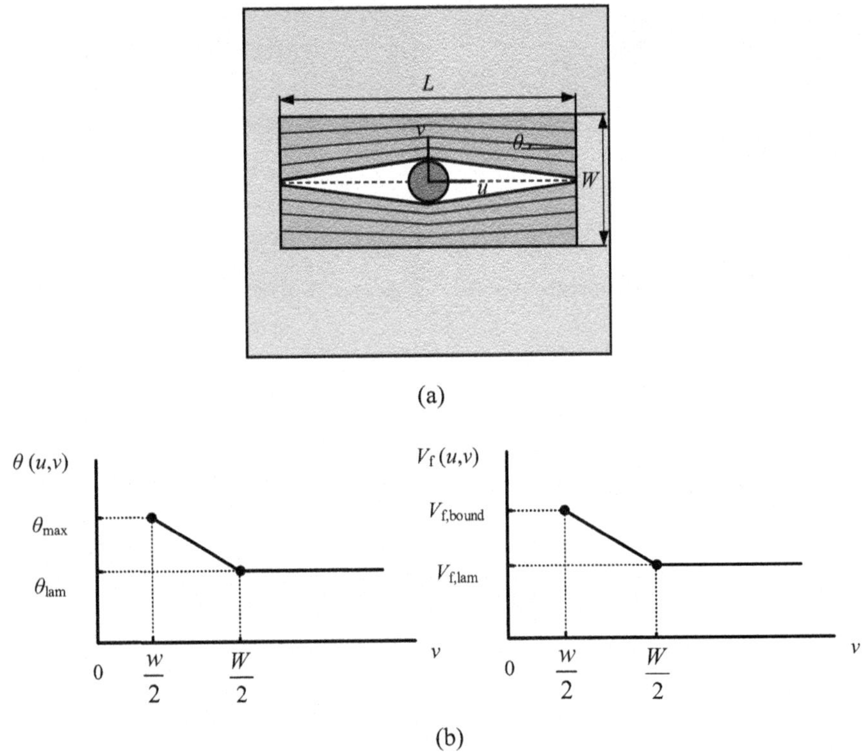


**Fig. 3.5** Mesoscopic geometrical RVE model. **a** A cross-section showing the modeling parameters, **b** assumed fiber distortion angle and fiber volume content in the cross-section

**Table 3.1** Geometrical parameters of the RVE for inserting thin and thick Z-pins

| | $d$ (mm) | $L$ (mm) | $W$ (mm) | $\theta_{max}$ (°) |
|---|---|---|---|---|
| Thin Z-pin | 0.28 | 2.0 | 0.8 | 4.0 |
| Thick Z-pin | 0.51 | 2.8 | 1.5 | 5.4 |

For the fiber-distorted zone, the modeling parameters are linearly varying between the two extreme values, as illustrated in Fig. 3.5b.

**Table 3.2** Curing kinetic parameters of 5228 epoxy resin

| Components | $A_1$ ($s^{-1}$) | $A_2$ ($s^{-1}$) | $n_1$ | $n_2$ | $m$ | $E_1$ (J/mol) | $E_2$ (J/mol) |
|---|---|---|---|---|---|---|---|
| Value | 338,000 | 6997 | 0.769 | 2.926 | 0.633 | 81,200 | 58,500 |

## 3.2 Modeling of the Cure-Dependent Properties

### *3.2.1 Cure Kinetics*

The cure of epoxy resin is a complex process that involves heat generation, thermal expansion and chemical shrinkage, which leads to variations of the phase of liquid mixtures into cross-linked solid structures. According to Lee et al. [7], the bimodal model, as a perfect autocatalytic model for epoxy resin, is derived to indicate the nonlinear relationship between curing rate with temperature and DoC during the curing process. Mathematically, the curing rate can be expressed as:

$$\frac{d\alpha}{dt} = K_1(T)(1-\alpha)^{n_1} + K_2(T)\alpha^m(1-\alpha)^{n_2} \tag{3.3}$$

where $m$, $n_1$ and $n_2$ are the reaction orders, $K_1(T)$ and $K_2(T)$ are the rate constants that depend on curing temperature governed by Arrhenius equations [8] as follows:

$$K_1(T) = A_1 \exp(-E_1/RT) \tag{3.4}$$

$$K_2(T) = A_2 \exp(-E_2/RT) \tag{3.5}$$

where $A_1$ and $A_2$ are the pre-exponential factors, $E_1$ and $E_2$ are the reaction activation energies, $T$ is the curing temperature, and $R$ is the universal gas constant. To obtain the curing kinetics parameters of the 5228 epoxy resin, a differential scanning calorimetry (DSC) test is conducted using DSC 200F3 of NETZSCH. Details of the DSC test can refer to the open literature [9]. From the DSC test results, the calibrated parameters of the kinetics model of 5228 epoxy resin are listed in Table 3.2.

### *3.2.2 Thermomechanical Properties of Resin Matrix*

Since epoxy resin transforms from the viscous liquid to the hard glassy solid in the cure, the thermomechanical properties are varied and can be defined by the evolution of DoC [10]. By using the mixing rule [11], the elastic modulus of the epoxy resin can be defined as:

$$E_{\mathrm{r}} = \begin{cases} E_{\mathrm{r}}^{0} & (\alpha < \alpha_{\mathrm{gel}}) \\ (1 - \alpha_{\mathrm{mod}})E_{\mathrm{r}}^{0} + \alpha_{\mathrm{mod}} E_{\mathrm{r}}^{\infty} & (\alpha \geq \alpha_{\mathrm{gel}}) \end{cases} \tag{3.6}$$

where the superscripts 0 and $\infty$ refer to initially uncured and finally cured states, respectively, $\alpha_{\mathrm{gel}}$ is the DoC at gelation point and is measured as 0.40 from the DSC test results, $\alpha_{\mathrm{mod}}$ is a function of DoC and $\alpha_{\mathrm{gel}}$ variations according to the following relation:

$$\alpha_{\mathrm{mod}} = \frac{\alpha - \alpha_{\mathrm{gel}}}{1 - \alpha_{\mathrm{gel}}} \tag{3.7}$$

In reference [12], the relationship of mechanical properties between the uncured state and the cured state of the epoxy resin can be denoted by:

$$\begin{cases} K_{\mathrm{r}}^{0} \approx K_{\mathrm{r}}^{\infty}/2.5 \\ G_{\mathrm{r}}^{0} \approx G_{\mathrm{r}}^{\infty}/100 \\ v_{\mathrm{r}}^{0} = (3K_{\mathrm{r}}^{0} - 2G_{\mathrm{r}}^{0})/(6K_{\mathrm{r}}^{0} + 2G_{\mathrm{r}}^{0}) \\ E_{\mathrm{r}}^{0} = 2(1 + v_{\mathrm{r}}^{0})G_{\mathrm{r}}^{0} \end{cases} \tag{3.8}$$

where $K$ and $G$ are the bulk and shear moduli, respectively, $v$ is the Poisson's ratio.

### 3.2.3 *Effective Properties of CFRP*

The mechanical properties of the composite can be stated in terms of the elastic constants of fiber and resin matrix using the micromechanics model [13, 14]. In the following equations, the subscripts 1, 2, and 3 represent three principal directions of the composite, r and f refer to the resin matrix and fiber, respectively. $V_{\mathrm{f}}$ is the fiber volume content of the composite. All the mechanical properties in 2 and 3 principal directions are interchangeable on account of the transverse isotropic characteristic of the composite. Thus, the elastic and shear moduli in the three principal directions respectively are expressed as follows:

$$\begin{cases} E_{11} = E_{\mathrm{f1}} V_{\mathrm{f}} + E_{\mathrm{r}}(1 - V_{\mathrm{f}}) + \left[\frac{4(v_{\mathrm{r}} - v_{\mathrm{f12}}^{2})K_{\mathrm{f}}K_{\mathrm{r}}G_{\mathrm{r}}(1 - V_{\mathrm{f}})V_{\mathrm{f}}}{(K_{\mathrm{f}} + G_{\mathrm{r}})K_{\mathrm{r}} + (K_{\mathrm{f}} - K_{\mathrm{r}})G_{\mathrm{r}}V_{\mathrm{f}}}\right] \\ E_{22} = E_{33} = \frac{1}{\frac{1}{4K} + \frac{1}{4G_{23}} + \frac{v_{12}^{2}}{E_{11}}} \\ G_{12} = G_{13} = G_{\mathrm{r}}\left[\frac{(G_{\mathrm{f12}} + G_{\mathrm{r}}) + (G_{\mathrm{f12}} - G_{\mathrm{r}})V_{\mathrm{f}}}{(G_{\mathrm{f12}} + G_{\mathrm{r}}) - (G_{\mathrm{f12}} - G_{\mathrm{r}})V_{\mathrm{f}}}\right] \\ G_{23} = G_{\mathrm{r}}\left[\frac{K_{\mathrm{r}}(G_{\mathrm{f23}} + G_{\mathrm{r}}) + 2G_{\mathrm{r}}G_{\mathrm{f23}} + K_{\mathrm{r}}(G_{\mathrm{f23}} - G_{\mathrm{r}})V_{\mathrm{f}}}{K_{\mathrm{r}}(G_{\mathrm{f23}} + G_{\mathrm{r}}) + 2G_{\mathrm{r}}G_{\mathrm{f23}} - (K_{\mathrm{r}} + 2G_{\mathrm{r}})(G_{\mathrm{f23}} - G_{\mathrm{r}})V_{\mathrm{f}}}\right] \end{cases} \tag{3.9}$$

The major and transverse Poisson's ratios respectively are given by:

$$\begin{cases} \nu_{12} = \nu_{13} = \nu_{f12} V_f + \nu_r (1 - V_f) + \left[ \frac{(\nu_r - \nu_{f12})(K_r - K_f) G_r (1 - V_f) V_f}{(K_f + G_r) K_r + (K_f - K_r) G_r V_f} \right] \\ \nu_{23} = \frac{2E_{11} K - E_{11} E_{22} - 4\nu_{12}^2 E_{22} K}{2E_{11} K} \end{cases} \tag{3.10}$$

The bulk moduli can be defined as:

$$\begin{cases} K = \frac{K_r (K_f + G_r) + G_r V_f (K_f - K_r)}{(K_f + G_r) - V_f (K_f - K_r)} \\ K_r = \frac{E_r}{2(1 - \nu_r - 2\nu_r^2)} \\ K_f = \frac{E_f}{3(1 - 2\nu_f)} \end{cases} \tag{3.11}$$

Following the idea of [15, 16], the thermal strain $\varepsilon^t$ can be expressed by the linear relation involving the variation of coefficient of thermal expansion (CTE) as follows:

$$\varepsilon^t = \beta \cdot T \tag{3.12}$$

Similarly, considering the perfect mixtures of viscous liquid and glassy solid states of the resin, the definition of CTE over the cure can be weighted by the DoC and determined as:

$$\beta_r = (1 - \alpha)\beta_r^u + \alpha \beta_r^c \tag{3.13}$$

where $\beta_r{}^u$ and $\beta_r{}^c$ stand for the CTEs of uncured and cured resin, respectively. The integrated CTEs of the composite can be also expressed using the micromechanics model as follows:

$$\beta_1 = \frac{\beta_{f1} E_{f1} V_f + \beta_r E_r (1 - V_f)}{E_{f1} V_f + E_r (1 - V_f)} \tag{3.14}$$

$$\begin{aligned} \beta_2 = \beta_3 = {} & (\beta_{f2} + \nu_{f12} \beta_{f1}) V_f + (\beta_r + \nu_r \beta_r)(1 - V_f) \\ & - [\nu_{f12} V_f + \nu_r (1 - V_f)] \beta_1 \end{aligned} \tag{3.15}$$

Chemical shrinkage of the resin mainly occurs after the gelation ($\alpha = 0.40$) and exhibits a linear dependence with DoC in the cure [17]. The chemical shrinkage strain $\varepsilon^c$ involving the coefficient of chemical shrinkage (CCS) can be determined by:

$$\varepsilon^c = \gamma \cdot \alpha \tag{3.16}$$

In this work, the chemical shrinkage of the composite can be calculated assuming the fibers as an inert material, and thus the shrinkage of the resin matrix can dominant the overall shrinkage behavior. Accordingly, the integrated CCSs of the composite in the three principal directions are respectively presented as follows:

**Table 3.3** Modeling parameters of T300/5228 base laminate

| Components | Categories | | |
|---|---|---|---|
| Fiber [18] | Density (kg/m$^3$) | $\rho_f = 1760$ | |
| | Young's modulus (MPa) | $E_{f1} = 230{,}000$ | $E_{f2} = E_{f3} = 15{,}000$ |
| | Shear modulus (MPa) | $G_{f12} = G_{f13} = 27{,}000$ | $G_{f23} = 7000$ |
| | Poisson's ratio | $\nu_{f12} = \nu_{f13} = 0.013$ | $\nu_{f23} = 0.03$ |
| | Coefficient of thermal expansion (°C$^{-1}$) | $\beta_{f1} = -0.7 \times 10^{-6}$ | $\beta_{f2} = \beta_{f3} = 12 \times 10^{-6}$ |
| Resin | Density (kg/m$^3$) | $\rho_r = 1340$ | |
| | Young's modulus (MPa) | $E_r^{\,0} = 20.8$ | $E_r^{\,\infty} = 3483$ |
| | Poisson's ratio | $\nu_r^{\,0} = 0.4977$ | $\nu_r^{\,\infty} = 0.3360$ |
| | Coefficient of thermal expansion (°C$^{-1}$) | $\beta_r^{\,u} = 88.4 \times 10^{-6}$ | $\beta_r^{\,c} = 35.36 \times 10^{-6}$ |
| | Coefficient of chemical shrinkage | $\gamma_r = -0.022$ | |

**Table 3.4** Modeling parameters of T300/BMI Z-pin [1]

| Parameters | $E_{11}$ (MPa) | $E_{22} = E_{33}$ (MPa) | $G_{12} = G_{13}$ (MPa) | $G_{23}$ (MPa) | $\nu_{12} = \nu_{13}$ | $\nu_{23}$ |
|---|---|---|---|---|---|---|
| Values | $1 \times 10^5$ | $3 \times 10^4$ | 4600 | 3900 | 0.35 | 0.4 |

$$\gamma_1 = \left[\frac{E_r(1 - V_f)}{E_{f1}V_f + E_r(1 - V_f)}\right]\gamma_r \tag{3.17}$$

$$\gamma_2 = \gamma_3 = (1 + \nu_r)(1 - V_f)\gamma - [\nu_{f12}V_f + \nu_r(1 - V_f)]\gamma_1 \tag{3.18}$$

Since Z-pins are the fiber-dominant material, the CTE and the CCS are both omitted in this work. To summarize, the material properties used in the model are given in Table 3.3 for the base laminate and Table 3.4 for the Z-pin, respectively.

### 3.2.4 *Curing Residual Stress Calculation*

Accounting for the evolutions of thermal expansion and chemical shrinkage during cure, the equivalent stresses can be defined as follows:

$$\{\sigma\} = [\mathbf{C}] \cdot \{\varepsilon^m\} = [\mathbf{C}] \cdot \{\varepsilon - \varepsilon^t - \varepsilon^c\} \tag{3.19}$$

where $\varepsilon$ is the total strain, $\varepsilon^m$, $\varepsilon^t$ and $\varepsilon^c$ stand for the mechanical, thermal and chemical shrinkage strains, respectively, **C** is the stiffness matrix.

## 3.3 Modeling of the Cure-Induced Interfacial Cracking

The constitutive response of cohesive sections is employed to characterize the interfacial behavior in terms of a bi-linear traction–separation law. As shown in Fig. 3.6, a linear elastic ascending phase is applied for the un-cracked elastic response, and a linear damage law is introduced for the cracking evolution. According to Simonovski and Cizelj [19], the definition of the interfacial traction stresses ($\sigma_n$, $\sigma_s$ and $\sigma_t$) and their corresponding displacements ($\delta_n$, $\delta_s$ and $\delta_t$) can be obtained by:

$$\begin{Bmatrix} \sigma_n \\ \sigma_s \\ \sigma_t \end{Bmatrix} = \begin{Bmatrix} [1-D(\delta)]K_{nn} & 0 & 0 \\ 0 & [1-D(\delta)]K_{ss} & 0 \\ 0 & 0 & [1-D(\delta)]K_{tt} \end{Bmatrix} \begin{Bmatrix} \delta_n \\ \delta_s \\ \delta_t \end{Bmatrix} \tag{3.20}$$

where the subscripts n, s and t denote the normal and two orthogonal shear directions, respectively, $K_{nn}$, $K_{ss}$ and $K_{tt}$ are the elastic stiffness in absence of cracking, $D(\delta)$ is a scalar index to characterize the cracking evolution.

Once the cracking begins, the stiffness of the interface is degraded depending on $D(\delta)$ which can be obtained by the following expression:

$$D(\delta) = \begin{Bmatrix} 0, & \delta < \delta^0 \\ \frac{\delta^f(\delta-\delta^0)}{\delta(\delta^f-\delta^0)}, & \delta^0 \le \delta \le \delta^f \end{Bmatrix} \delta = \sqrt{\langle\delta_n\rangle^2 + \delta_s^2 + \delta_t^2} \tag{3.21}$$

where $\delta^0$ and $\delta^f$ are the initial and ultimate failure displacements under mixed-mode traction stresses, respectively. The symbol $\langle\rangle$ is the Macaulay bracket with a definition of $\langle x\rangle = (x + |x|)/2$.

The QUADS law is available for calculating the beginning of the interfacial cracking.

$$\left\{\frac{\langle\sigma_n\rangle}{\sigma_n^0}\right\}^2 + \left\{\frac{\sigma_s}{\sigma_s^0}\right\}^2 + \left\{\frac{\sigma_t}{\sigma_t^0}\right\}^2 = 1 \tag{3.22}$$

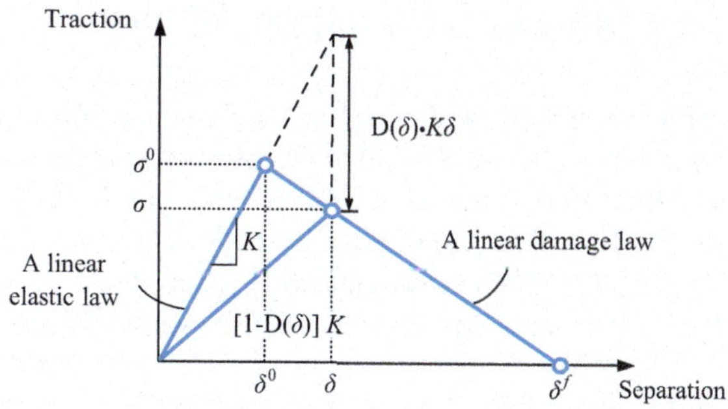


**Fig. 3.6** Bi-linear cohesive law for the interface region

**Table 3.5** Modeling parameters of the interface

| $K_{nn}$ (N/mm$^3$) | $K_{ss}$ (N/mm$^3$) | $K_{tt}$ (N/mm$^3$) | $\sigma_n{}^0$ (MPa) | $\sigma_s{}^0$ (MPa) |
|---|---|---|---|---|
| $1 \times 10^5$ | $1 \times 10^5$ | $1 \times 10^5$ | 60 | 90 |
| $\sigma_t{}^0$ (MPa) | $G_I{}^C$ (N/mm) | $G_{II}{}^C$ (N/mm) | $G_{III}{}^C$ (N/mm) | $\eta_{BK}$ |
| 90 | 0.21 | 0.78 | 0.78 | 1.94 |

The B-K law, acting as a mixed-mode energy criterion, is applied to estimate the completion of the interfacial cracking.

$$G^C = G_I^C + \left(G_{II}^C - G_I^C\right)\left(\frac{G_{II} + G_{III}}{G_I + G_{II} + G_{III}}\right)^{\eta_{BK}} \tag{3.23}$$

where $\eta_{BK}$ is the mixed-mode index, $G_I{}^C$, $G_{II}{}^C$, and $G_{III}{}^C$ represent the fracture toughness, $G_I$, $G_{II}$ and $G_{III}$ are introduced to compute the fracture energy in the process regarding as:

$$G_I = \int_0^{\delta_n} \sigma_n(\delta)\mathrm{d}\delta_n$$

$$G_{II} = \int_0^{\delta_s} \sigma_s(\delta)\mathrm{d}\delta_s$$

$$G_{III} = \int_0^{\delta_t} \sigma_t(\delta)\mathrm{d}\delta_t \tag{3.24}$$

The cohesive properties used in this work are given in Table 3.5 based on the open literature [20–23].

## 3.4 Finite Element Implementation

With the aim of predicting the interfacial failure mechanism of Z-pinned CFRP composites after the cure, an RVE model of a single Z-pinned laminate is established. The RVE model is created in ABAQUS according to the microstructural morphology (as shown in Fig. 3.3) of the Z-pinned specimen and presented in Fig. 3.7.

Since the Z-pin misalignment is reduced to 4.8° with few fiber crimping by the PHZ process, the Z-pin is assumed to be vertically inserted into the laminates without misalignment and fiber crimping. The diameter of the Z-pin is 0.51 mm. The length of the resin pocket is 3 mm. The thickness of each ply is 0.125 mm, with a stacking sequence of $[0/45/90/-45]_s$. A discrete layer of cohesive elements (COH3D8) is

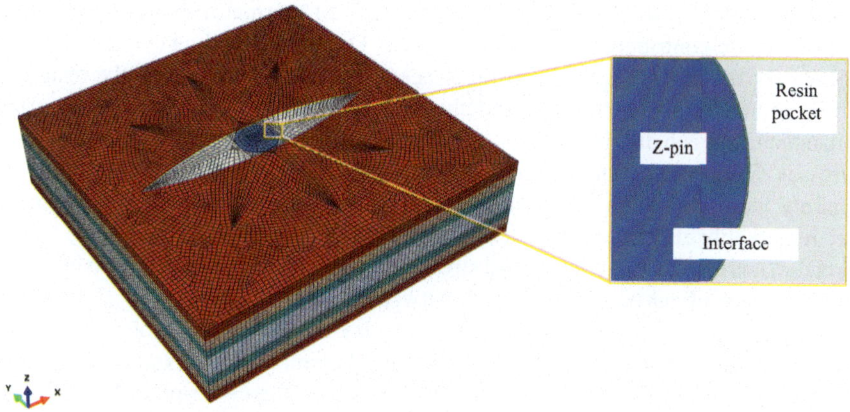


**Fig. 3.7** Finite element mesh of the RVE model

inserted between Z-pin and resin pocket with a thickness of 0.001 mm to simulate the interface. The mesh size is defined as 0.05 mm after conducting mesh sensitivity analysis to obtain high grid quality. The bottom surface of the RVE model is fully constrained while the other surfaces are free.

Two steps of simulations are conducted: firstly, a cure simulation is proposed for characterizing the curing residual stresses of the RVE model subjected to the curing process; then, the interfacial cracking behavior of the RVE model is simulated by combining the residual stresses as a predefined boundary. The QUADS coupled with the B-K criterion is unitized to estimate the specific location and amount of the interfacial cracking area.

## 3.5 Simulation Results

### *3.5.1 Curing Residual Stress Prediction*

The evolutions of DoC, temperature and curing residual stresses under the recommended curing profiles are given in Fig. 3.8. It can be seen in Fig. 3.8a that the DoC after the heating process reaches about 0.22 below the gelation point ($\alpha = 0.40$). Accordingly, the gelation and variation of resin properties will occur after the heating process. Then, the DoC is significantly improved in the holding process and ultimately maintains a constant value of 0.96 in the cooling process. The evolutions of the maximum curing residual stresses of each phase are described in Fig. 3.8b. Each stress plot can be systematically divided into three stages in terms of three different curing stages. During the heating stage of the cure (Stage I), almost no residual stresses occur. Since the resin is in the liquid state, the stress transmission among each phase is so weak, and the mismatch of thermal response has a minor impact on the contribution of residual stresses. During the holding stage of the cure (Stage II), the residual stresses begin to accumulate. It is known that the resin can show strong cross-link behavior and chemical shrinkage after gelation. The curing mismatch between different phases can be responsible for accumulating the residual stresses. During the cooling stage of the cure (stage III), the residual stresses are sharply increased. High cured modulus of the resin and further curing mismatch are remarkably effective at developing the residual stresses. At the end of the curing process, it is obvious that the highest stress concentration is produced in the interface due to the significant curing mismatch between the resin and Z-pin.

Consider the symmetry of the composite lay-up, the distributions of residual stresses in the first four layers of the model are shown in Fig. 3.9, indicative of four different ply orientations of the Z-pinned CFRP composites (− 45-ply, 90-ply, 45-ply,

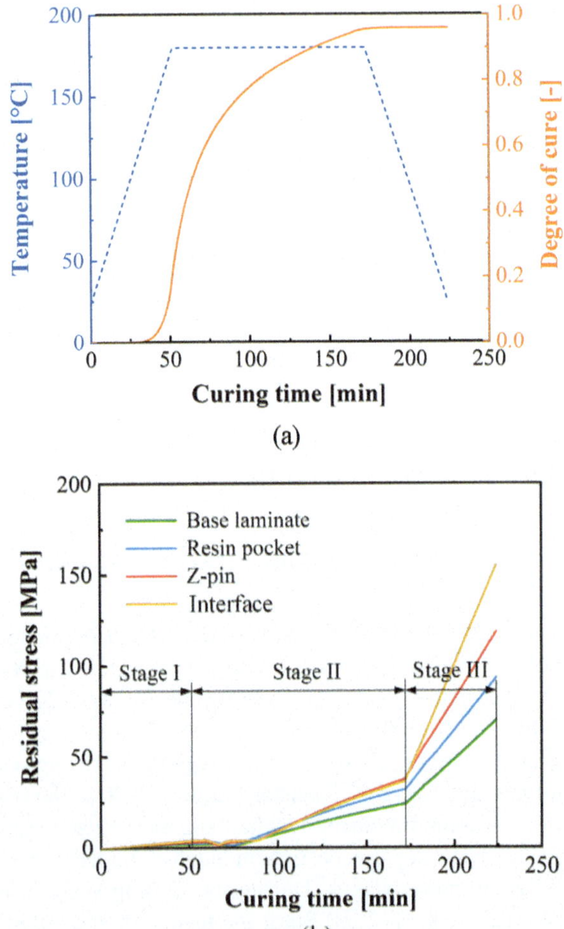


**Fig. 3.8** Illustrations of the RVE curing simulation. **a** DoC and temperature. **b** Residual stresses

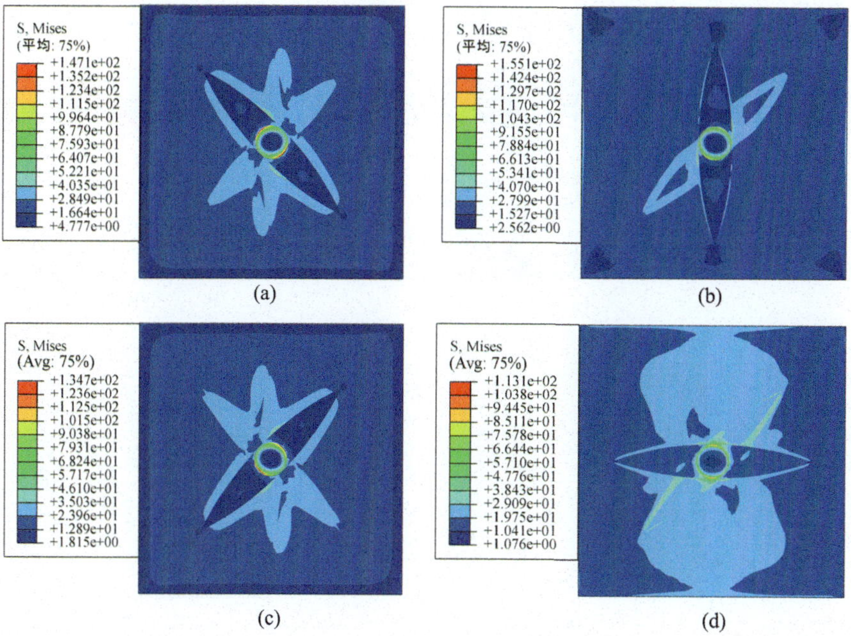


**Fig. 3.9** Distributions of residual stresses in different Z-pinned layers. **a** – 45-ply. **b** 90-ply. **c** 45-ply. **d** 0-ply

0-ply). Regardless of the various ply orientations, the highest stress concentration always occurs in the interface surrounding the Z-pin.

The distributions of maximum principal and normal stresses in the 45-ply of the model are further given in Fig. 3.10. As can be seen, the resin pocket generally experiences a tensile state, while the partial region nearby the Z-pin exhibits a compressive state. In terms of normal residual stresses, the resin pocket is completely under a tensile state in both longitudinal and two transverse directions. The base laminate is mostly under a compressive state in the longitudinal direction and a tensile state in the transverse direction. The Z-pin is mainly under a compressive state in both longitudinal and transverse in-plane directions.

The evolution of curing residual stresses in Z-pin is exhibited in Fig. 3.11. The stress concentration initiates in the segments that are facing the resin pocket and gradually extends to the other regions. The maximum residual stress is 118.60 MPa and appears in a helix path, due to the four different ply orientations of the base laminates through the Z-pin axis.

For the interface part, the detailed distributions of Mises residual stresses and maximum principal stresses are presented in Figs. 3.12 and 3.13, respectively. The ultimate Mises stress is estimated to be 155.1 MPa and the corresponding maximum principal stress is 85.29 MPa. The stress concentration occurs along the helix path, illustrating a partially non-uniform force state of the interface.

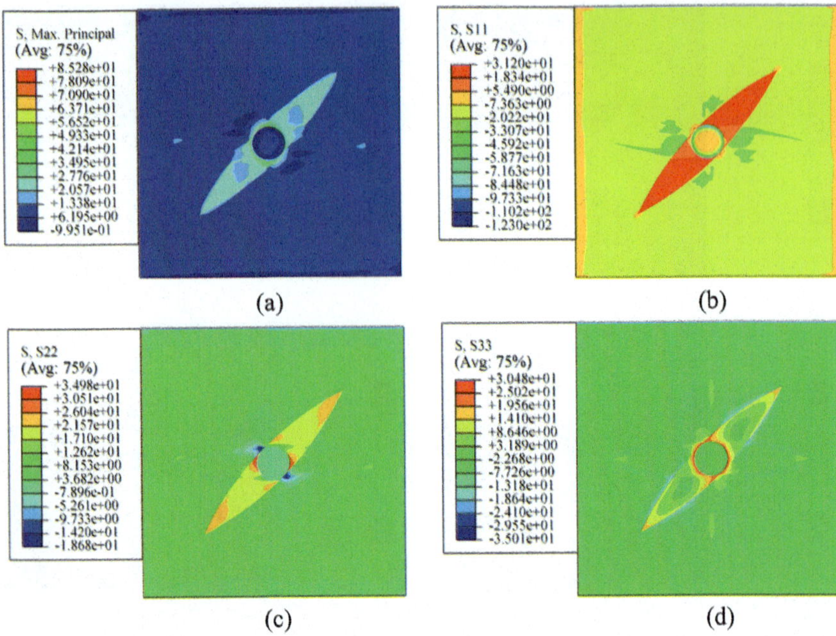


**Fig. 3.10** Distributions of residual stresses in the 45-ply Z-pinned layer. **a** Maximum principal stress. **b** Longitudinal stress. **c** Transverse in-plane stress. **d** Transverse out-of-plane stress

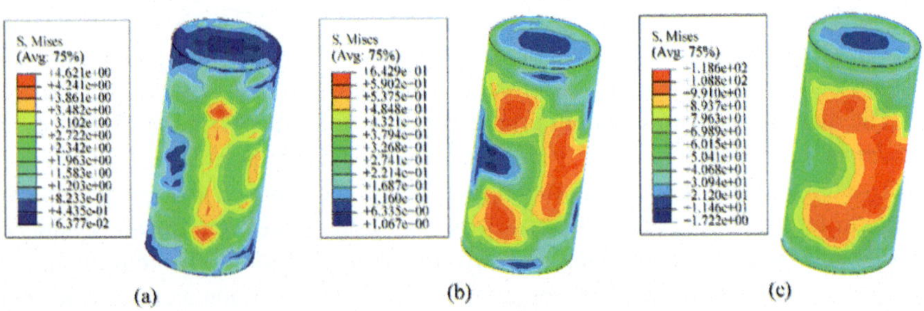


**Fig. 3.11** Evolution of Mises residual stresses in the Z-pin region. **a** Initial stress. **b** Accumulated stress. **c** Ultimate stress

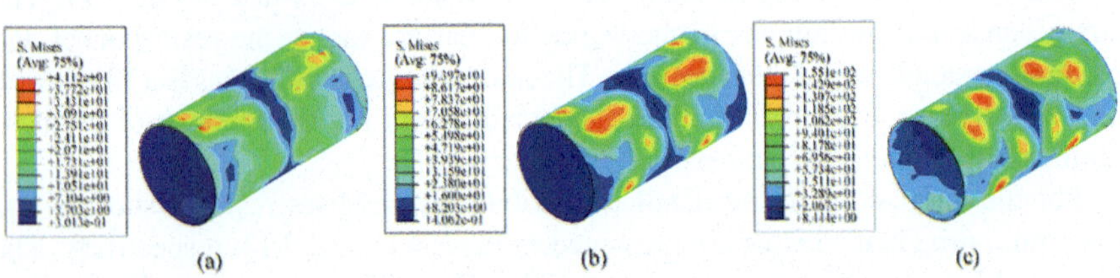


**Fig. 3.12** Evolution of Mises residual stresses in the interface region. **a** Initial stress. **b** Accumulated stress. **c** Ultimate stress

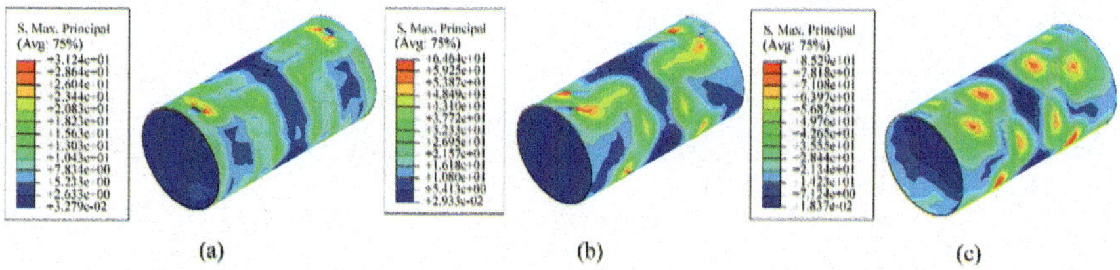


**Fig. 3.13** Evolution of maximum principal residual stresses in the interface region. **a** Initial stress. **b** Accumulated stress. **c** Ultimate stress

## 3.5.2 *Interfacial Cracking Evolution*

The effect of residual stresses on the interfacial cracking is analyzed. Both initial damage index (QUADSCRT) and complete failure index (SDEG) are applied to examine the failure progression. It is depicted in Fig. 3.14 that the interface is significantly weakened after cure. The crack location starts in the segments corresponding to the high stress concentration and gradually progresses to the other affected region. Debonding area ratio is the percentage of cracking area (SDEG = 1) in the total area. The estimated debonding area ratio is about 40% at the end of the cure.

The effect of interfacial strength on the debonding behavior is also examined. For comparison, a lower interfacial strength ($\sigma_n{}^0 = \sigma_s{}^0 = 30$ MPa, $\sigma_t{}^0 = 45$ MPa) is also selected to evaluate the interfacial cracking behavior. The evolution of interfacial cracking is depicted in Fig. 3.15. It is found that more initial damage areas appear owing to the low interfacial strength. The damage propagation path is similar as

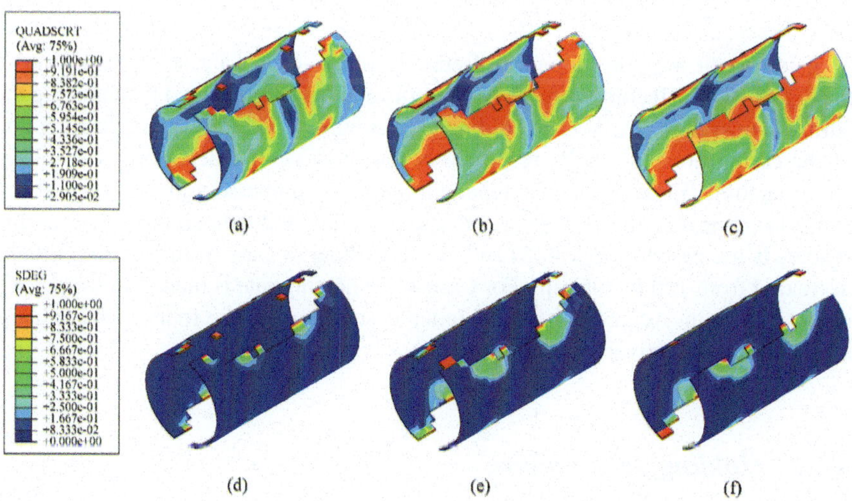


**Fig. 3.14** Evolution of failure mechanism in the interface region. **a–c** QUADSCRT results. **d–f** SDEG results

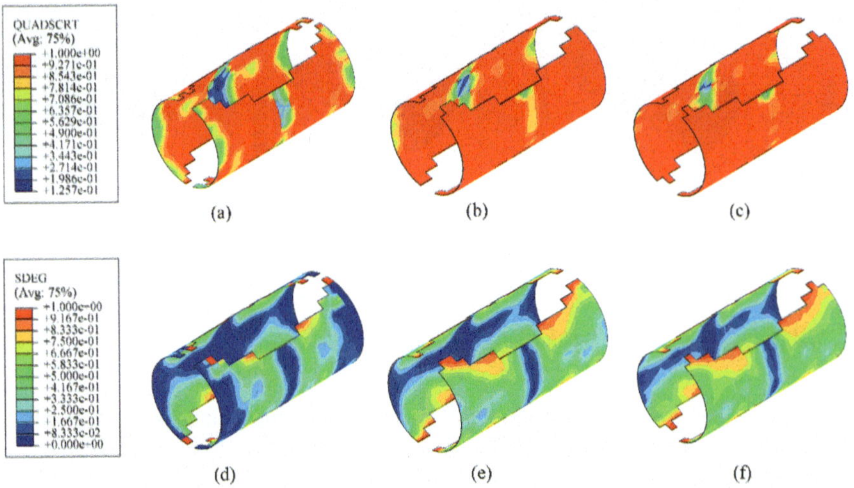


**Fig. 3.15** Effect of the low interfacial strength on the interfacial cracking. **a–c** QUADSCRT results. **d–f** SDEG results

described before. The ultimate debonding area ratio is increased to 43%. The interfacial cracking in the RVE model and the microstructural observations are compared in Fig. 3.16.

## 3.6 Influence of Curing Process on the Interfacial Cracking

To gain insight into the influence of curing process on the interfacial cracking of the Z-pinned CFRP composites, a parametric analysis is conducted by changing the curing temperature profile. The holding temperature, holding time, and heating rate of the curing process can be respectively regarded as an independent variable in the simulation. Three designed temperature profiles compared with the original one (Case 1) are presented in Fig. 3.17. In Case 2, the holding temperature of the curing process is increased from 180 to 186 °C. Regarding to Case 3, the holding time is decreased from 120 to 105 min. For Case 4, the heating rate is increased from 3 °C/min to 6 °C/min. Accordingly, the detailed descriptions of the four cases are listed in Table 3.6, respectively.

### 3.6.1 *Holding Temperature*

Compared with Case 1, the evolutions of DoC and the ultimate residual stresses in Case 2 are displayed in Fig. 3.18. It can be seen in Fig. 3.18a that the DoC of Case 2

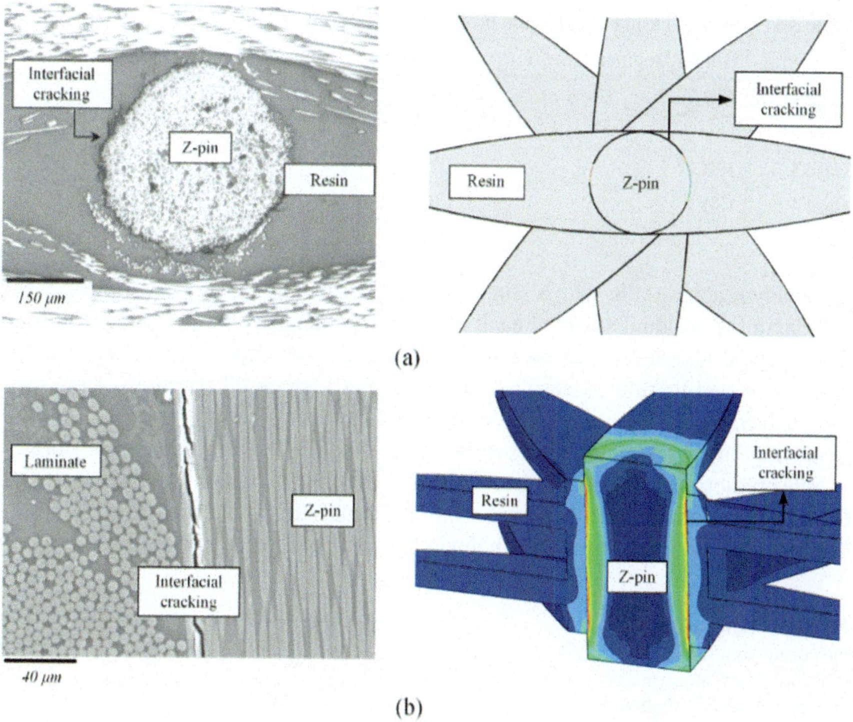


**Fig. 3.16** Comparisons of experiments and simulation results of partially debonding. **a** In a cross-sectional view. **b** In a side-sectional view

**Fig. 3.17** Designed curing temperature profiles in different cases

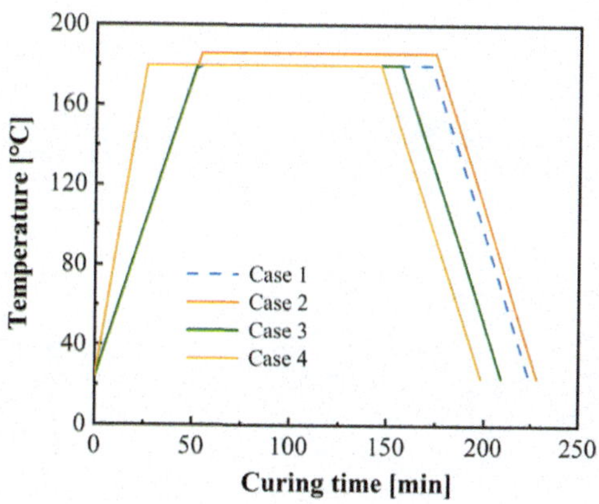

**Table 3.6** Designed curing temperature profiles

| | Holding temperature (°C) | Holding time (min) | Heating rate (°C/min) |
|---|---|---|---|
| Case 1 | 180 | 120 | 3 |
| Case 2 | 186 | 120 | 3 |
| Case 3 | 180 | 105 | 3 |
| Case 4 | 180 | 120 | 6 |

ultimately reaches 0.98, with a slight increase in the holding process. In Fig. 3.18b, the maximum residual stress of each phase is significantly increased, especially for the interface region. It is known that the increased holding temperature can lead to an increased thermal release together with the high cured modulus of the resin. This behavior is detrimental for the modulus matching between phases, and thus producing larger residual stresses of the Z-pinned CFRP composites in the cure.

The predictions of interfacial residual stresses together with interfacial cracking in Case 2 are given in Fig. 3.19. The maximum residual stress is increased to 177.3 MPa. The corresponding maximum principal stress is about 99.11 MPa. They are both

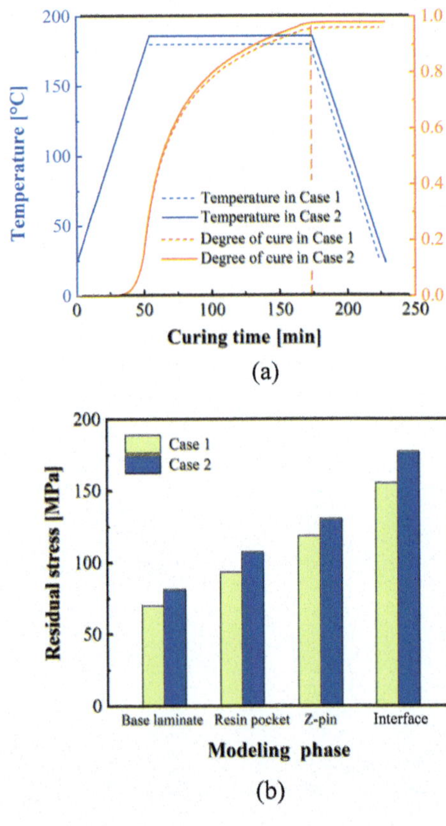


**Fig. 3.18** Comparisons of the curing simulation results in Case 1 and Case 2. **a** DoC and temperature. **b** Residual stresses

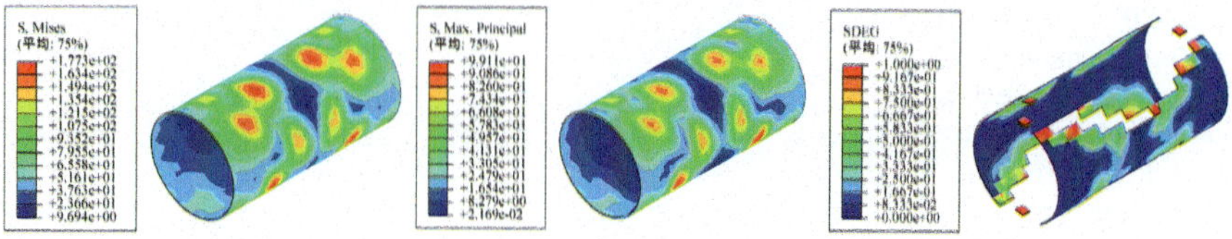


(a) (b) (c)

**Fig. 3.19** Simulation results of the interface region under Case 2. **a** Mises residual stress. **b** Maximum principal residual stress. **c** Ultimate interfacial cracking

remarkably increased compared with those of Case 1. Consequently, the cracking progresses to more area compared with Case 1. Debonding area ratio in Case 2 is estimated to be 66% with an obvious increase.

### *3.6.2 Holding Time*

With the comparison of Case 1, the estimated DoC and curing residual stresses in Case 3 are both shown in Fig. 3.20. As presented in Fig. 3.20a, when the holding process is finished, the DoC in two cases begins to appear difference. In the absence of holding time in Case 3, the DoC stops improving and maintains the constant value of 0.935 until the curing process is completed. The curing residual stresses of each phase are also limited in Case 3, as shown in Fig. 3.20b. This is owing that the reduced holding time produces an incomplete curing reaction and low modulus of resin. Thus, the inhibition of the curing mismatch between different phases will alleviate the partial stress concentration and remarkably reduce the development of residual stresses.

The interfacial stress distribution and cracking in Case 3 are exhibited in Fig. 3.21. The maximum Mises stress is reduced to 143.8 MPa and the maximum principal stress is 81.84 MPa. Compared with Case 1, the interfacial cracking in Case 3 is also slightly limited, and the ultimate debonding area ratio is decreased to 38%.

### *3.6.3 Heating Rate*

Compared with Case 1, the simulated results involving DoC evolution and curing residual stresses in Case 4 are collected in Fig. 3.22. It can be seen in Fig. 3.22a that the DoC in Case 4 is improved rapidly as the heating rate increases. Then, the increase in the DoC gradually slows down and finally reaches 0.95 (0.96 in Case 1). As shown in Fig. 3.22b, the maximum residual stress of each phase is increased. This is because the increased heating rate can shorten the curing time, and make the

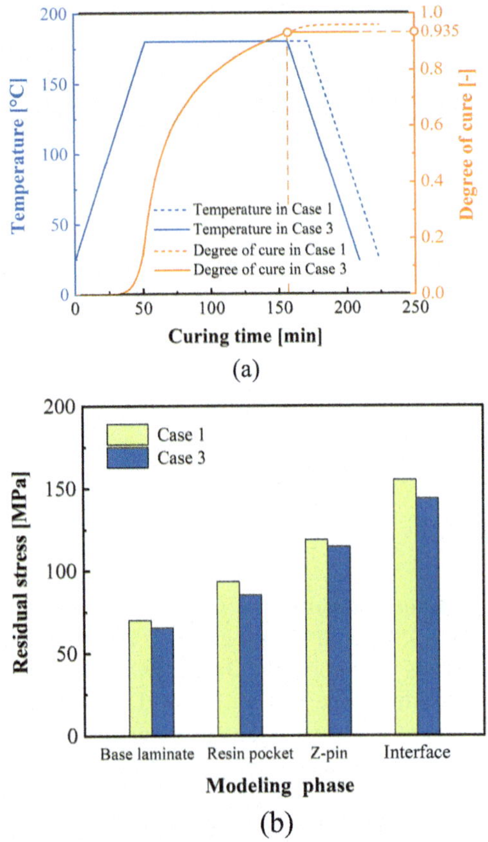


**Fig. 3.20** Comparisons of the curing simulation results in Case 1 and Case 3. **a** DoC and temperature. **b** Residual stresses

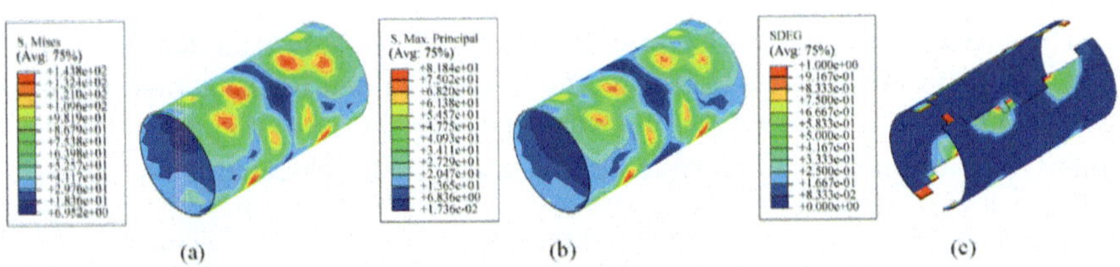


**Fig. 3.21** Simulation results of the interface region under Case 3. **a** Mises residual stress. **b** Maximum principal residual stress. **c** Ultimate interfacial cracking

resin flow insufficiently during the curing process. Subsequently, the non-uniform curing can explicitly result in the increased residual stresses.

The simulated results of interfacial residual stresses together with interfacial cracking in Case 4 are shown in Fig. 3.23. The highest stress concentration is extremely increased to 168.0 MPa with the corresponding maximum principal stress

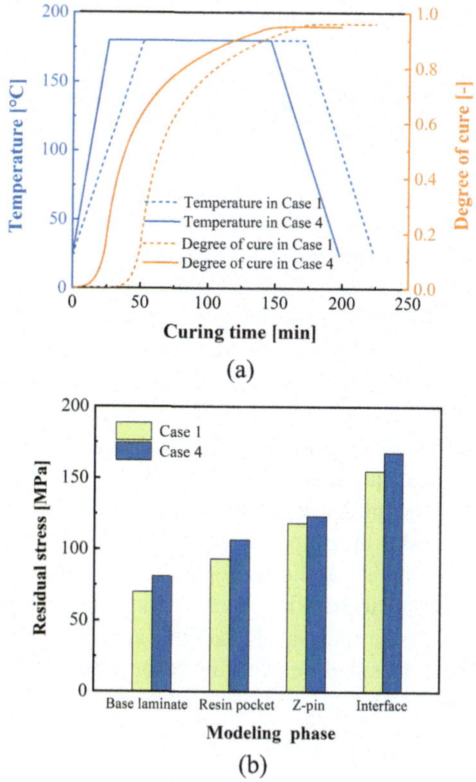


**Fig. 3.22** Comparisons of the curing simulation results in Case 1 and Case 4. **a** DoC and temperature. **b** Residual stresses

of 92.99 MPa. The ultimate debonding area ratio is increased to 62%, showing more cracking propagation of the interface.

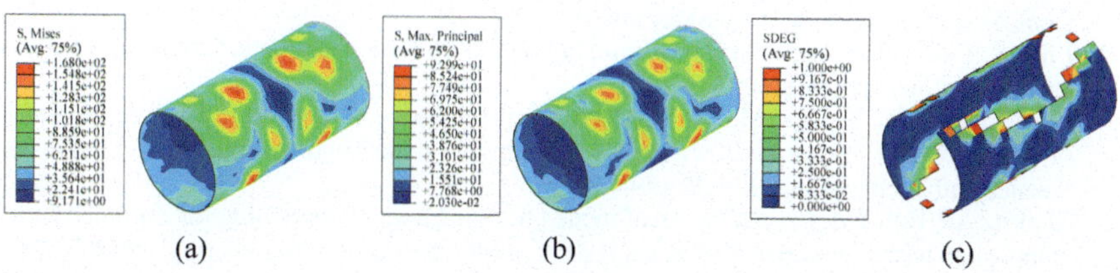


**Fig. 3.23** Simulation results of the interface region under Case 4. **a** Mises residual stress. **b** Maximum principal residual stress. **c** Ultimate interfacial cracking

## 3.7 Conclusions

In this chapter, a micro-mechanical modeling method is introduced for analyzing the evolution of cure-induced residual stresses and its effect on the interfacial cracking of Z-pins. Variation of resin properties and chemical shrinkage in the curing process are both taken into consideration in the simulation. The distributions of curing residual stresses in Z-pinned CFRP composites are predicted. The specific location and amount of interfacial cracking caused by the curing residual stresses are also estimated.

The stress concentration mainly occurs in the Z-pin/resin interface owing to the huge mismatch of thermal expansion and chemical shrinkage between Z-pin and resin pocket. The local residual stresses can exceed the strength of the interface. Therefore, partially interfacial cracking occurs around the Z-pin. This simulation result is confirmed by the experimental observations of microstructure of the Z-pinned specimens.

The influences of curing process on the interfacial cracking are also investigated via using the proposed approach. Influences of holding temperature, holding time and heating rate of the curing process are discussed, respectively. The most pronounced interfacial cracking comes from the increased holding temperature of the cure. Since the interfacial cracking around Z-pins is an inherent problem, comprehensive optimization of curing parameters for reducing interfacial cracking is extremely necessary.

## References


1. Bianchi F, Zhang X. A cohesive zone model for predicting delamination suppression in Z-pinned CFRP composites. Composites Science and Technology. 2011;71:1898–907.
2. Lin Y. Role of matrix resin in delamination onset and growth in composite laminates. Composites Science and Technology. 1988;33:257–77.
3. Liu KS, Tsai SW. A progressive quadratic failure criterion for a laminate. Composites Science and Technology. 1998;58:1023–32.
4. Benzeggagh ML, Kenane M. Measurement of mixed-mode delamination fracture toughness of unidirectional glass/epoxy composite with mixed-mode bending apparatus. Composites Science and Technology. 1996;56:439–49.
5. Mouritz AP, Koh TM. Re-evaluation of mode I bridging traction modelling for Z-pinned CFRP composites based on experimental analysis. Composites Part B: Engineering. 2014;56:797–807.
6. Pierreux G, Wu L, Van Hemelrijck D, Massart T. Evaluation of microdamage initiation in Z-pinned CFRP composites by means of automated RVE computations. Composite Structures. 2018;206:104–15.
7. Lee SN, Chiu MT, Lin HS. Kinetic model for the curing reaction of a tetraglycidyl diamino diphenyl methane/diamino diphenyl sulfone (TGDDM/DDS) epoxy resin system. Polymer Engineering and Science. 1992;32(15):1037–46.
8. Logan SR. The origin and status of the Arrhenius equation. Journal Of Chemical Education. 1982;59 (4):279.
9. Kamal MR. Thermoset characterization for moldability analysis. Polymer Engineering and Science. 1974;14(3):231–9.

10. Bogetti TA, John J GW. Process-induced stress and deformation in thick-section thermoset composite laminates. Journal of composite materials. 1992;26:626–60.
11. Johnston AA. An integrated model of the development of process-induced deformation in autoclave processing of composite structures: The University of New Brunswick, 1992.
12. Svanberg JM, Holmberg JA. Prediction of shape distortions. Part II. Experimental validation and analysis of boundary conditions. Composites Part A: Applied Science and Manufacturing. 2004;35:723–34.
13. Whitney JM. Elastic Moduli of Unidirectional Composites with Anisotropic Filaments. 1967;1:188–93.
14. Krajcinovic D. Micromechanically Inspired Continuous Models of Brittle Damage. In: Lemaitre J, editor. Handbook of Materials Behavior Models. Burlington: Academic Press; 2001. p: 417–20.
15. Abou Msallem Y, Jacquemin F, Boyard N, Poitou A, Delaunay D, Chatel S. Material characterization and residual stresses simulation during the manufacturing process of epoxy matrix composite. Composites Part A: Applied Science and Manufacturing. 2010;41:108–15.
16. Rabearison N, Jochum C, Grandidier JC. A FEM coupling model for properties prediction during the curing of an epoxy matrix. Computational Materials Science. 2009;45:715–24.
17. Kravchenko OG, Kravchenko SG, Pipes RB. Chemical and thermal shrinkage in thermosetting prepreg. Composites Part A: Applied Science and Manufacturing. 2016;80:72–81.
18. Jia Y, Yan W, Liu HY. Carbon fibre pullout under the influence of residual thermal stresses in polymer matrix composite. Computational Materials Science. 2012;62:79–86.
19. Simonovski I, Cizelj L. Cohesive zone modeling of intergranular cracking in polycrystalline aggregates. Nuclear Engineering and Design. 2015;283:139–47.
20. Zhang B, Allegri G, Yasaee M, Hallett SR. Micro-mechanical finite element analysis of Z-pins under mixed-mode loading. Composites Part A: Applied Science and Manufacturing. 2015;78: 424–35.
21. Melro Antonio R, Serra J, Allegri G, Hallett SR. An energy-equivalent bridging map formulation for modelling delamination in through-thickness reinforced composite laminates. International Journal of Solids and Structures. 2020;202:153–65.
22. Koyanagi J, Yoshimura A, Kawada H, Aoki Y. A numerical simulation of time-dependent interface failure under shear and compressive loads in single-fiber composite. Applied Composite Materials. 2010;17(1):31–41.
23. Koyanagi J, Ogihara S, Nakatani H, Okabe T, Yoneyama S. Mechanical properties of fiber/matrix interface in polymer matrix composite. Advanced Composites and Hybrid Materials. 2014; 23(5-6):551–70.

# Chapter 4
# Multiscale Modeling of the Mechanical Properties of Z-pinned CFRP Composites

Composites show the hierarchical structures in nature and every constituent affects the overall mechanical behaviours. The Z-pinning process leads to more complex microstructures in these composites, including the resin-rich regions and fiber unevenness. Furthermore, the curing process generates thermal residual stresses in Z-pinned composites, leading to partial or complete cracking along Z-pin/composite interfaces. These complex microstructures and stress distribution inevitably change the mechanical properties of Z-pinned composites. However, due to the tremendous computation cost, little effort has been devoted to the investigation into the effects of the microstructures on the mechanical properties of Z-pinned composites.

In this chapter, we developed a multiscale framework to evaluate the delamination behaviours of Z-pinned composites considering the effects of Z-pin/composite interfacial cracks. Unit cell models are used to represent the typical microstructures and micro-mechanisms of Z-pinned composites with diverse interfacial conditions and to predict Z-pin pullout behaviours. The derived Z-pin bridging laws are then used to simulate the behaviours of individual Z-pins in macroscale models. The load-versus-displacement relationships of Z-pinned composites under mode-I loading conditions are predicted using macroscale models and then verified experimentally. A parametric study reveals that interfacial cracks can affect the failure behaviours of Z-pinned composites by changing the microscale bridging mechanisms of individual Z-pins. It is theoretically feasible to customise Z-pin/composite interfacial bonding properties, Z-pin surface roughness, and curing temperature to optimise the delamination properties of Z-pinned composites.



Y. Xu and W. Zhang, *Modeling of Z-pinned Carbon Fiber-Reinforced Polymer (CFRP) Composite*, https://doi.org/10.1007/978-981-96-3628-0_4

## 4.1 Multiscale Simulation Methodology of Z-pinned CFRP Composites

### *4.1.1 Simulation Strategy of Z-pinned CFRP Composites*

Multiscale modelling has been typically applied to analyze Z-pinned composites laminates from multiple scales. The behaviors of individual Z-pin could be characterized in micro-scale simulations, which provide detailed insight into the micromechanics and micro-structure of Z-pinned composites laminates. The high accuracy of micro-scale modelling is achieved at the cost of computation time. That is, discrete modelling of every single Z-pin, pin-laminate interface and surrounding laminates in detail is impractical for the macro-scale cases with numerous Z-pins due to the prohibitive computation cost. On the other hand, behaviors of large numbers of Z-pins and composite continuum can be simulated in macro-scale models, where the bridging tractions of individual Z-pins are reasonably idealized according to the traction–separation laws. The mechanical properties of Z-pinned composites and the macroscopic effects of Z-pinning parameters can be predicted by these macro-scale models, although some microscale details have to be ignored. Combing the micro- and macro-scale analysis through the transmission of Z-pin bridging laws, the multi-scale simulation provides a better choice with the compromise of fidelity and efficiency.

Most multi-scale research of Z-pinned composites possesses a hierarchical bottom-up strategy, where the Z-pin traction–separation laws can be derived from the micro-scale models and further unidirectionally passed into the macro-scale models. Micro-scale models depict the bridging and failure behaviors of individual Z-pins and macro-scale models predict the macroscopic mechanical properties of Z-pinned composites subjected to various loading conditions. On the contrary, information in other composite analyses could flow in bottom-up, top-down or concurrent sequences, among more scales ranging from micro-, meso- to macro-scales. Additionally, homogenization has also been found in these simulation analyses of composite without Z-pins [1, 2]. More information about multi-scale methods for composite could be found in the reference [3, 4].

Z-pinned composites with periodic structures can be analyzed by Representative Volume Element (RVE) or Unit-Cell (UC) models. These micro-scale models center on the smallest unit containing a single Z-pin, pin-laminate interface, surrounding matrix and other microstructures. The mechanisms, such as elastic deformation, debonding, frictional sliding, rupture and splitting of Z-pins, Z-pin pullout and fiber failure, have been simulated. The contribution of various mechanisms to overall Z-pin behaviors has also been assessed, complementing experimental observations. Insightful Z-pin traction–separation laws are obtained.

To achieve an accurate simulation of microstructures and elaborate analysis of micromechanisms of Z-pinning, the aforementioned characteristics are simulated in detail. Fiber waviness and resin-rich regions are assumed to be restricted in a rectangle region, while remote laminates are regarded to be unaffected. One approach to explicitly simulate fiber waviness and resin-rich region is the star-like ply-level mesh method [5]. Resin-rich regions are defined by separate assignments of the resin matrix and laminate properties. Multiple piles with various orientations are stacked together to generate a star-like mesh. Fiber waviness is regarded to be sinusoidal around the Z-pin; therefore, fiber waviness angle can be determined by the maximum fiber waviness angle and Z-pin position. The maximum fiber waviness angle is assumed to occur on both edges of the resin-rich region, which is controlled by Z-pin diameter and rectangle length. The out-of-plane crimp angle is always ignored in simulations, while laminate swelling is implicitly considered by increased laminate thickness. The resin matrix surrounding the Z-pin is assumed to be homogeneous and isotropic, and therefore the resin-rich region can be modeled as identical solid elements with corresponding material properties. The resin between Z-pin and laminates can be simplified to a tie constraint as well. It is also justified to profile the resin-rich zone and embedded Z-pin by a void with a refined mesh, since the resin-rich region with much lower strength and stiffness is highly susceptible to cracks under applied loading. Z-pin misalignment can be transformed to the vertical Z-pin under misaligned loading. Another method to simulate Z-pin misalignment is to extrude the created 2D mesh at a characteristic offset angle and manually mesh the empty space in the cuboid cell. Errors induced by abnormal elements in this model can be effectively solved by model verification procedures. Furthermore, the through-thickness and in-plane offset angles can be simultaneously modeled by translating the top surface nodes of a straight Z-pin. The interfacial bonding between Z-pin and laminates can be described by a layer of zero-thickness cohesive elements. After the interfacial debonding, the following frictional sliding between the Z-pin and laminates can be depicted by surface-to-surface contact elements with defined friction coefficients. Contact happens when the element surface penetrates any Z-pin element on the target surface. Different contact statuses are reflected by changing moduli of the contact elements [6]. The resin-rich region can be connected to the Z-pin via tie constraints [7], cohesive contact interactions coupled with global Coulomb contact or joint interface elements. The simulation of crushing and spalling in the resin under mixed-mode loading has been achieved by implementing smaller resin parts connected by cohesive interactions [7]. The snubbing process in the resin has been modeled by embedding zero-thickness cohesive elements in the resin-rich zone [8]. The splitting and rupturing process of Z-pins can be simulated by zero-thickness cohesive elements normal and parallel to the Z-pin axis. When Z-pin splitting occurs, the friction between the newly exposed surfaces can be defined by the Coulomb friction contact between adjacent splitting strands. The effects of post-cure cool-down can be simulated by imposing a temperature difference field on all the nodes before the delamination loading. The effects of thermal stress have also been taken into account by implementing an initial compressive pressure contact between the Z-pin and surrounding matrix or setting the friction coefficient value.

The damage initiation of elements in the Finite Element models can be determined according to the quadratic nominal stress criterion and Weibull's failure criterion. In addition, the Benzeggagh and Kenane criterion (B-K law), power-law criterion and linear energy criterion are also applied based on the energy release rate, introducing less troublesome singularity. Critical strain energy release rates and empirical parameters in these criteria can be concluded from test results.

Periodic boundary conditions are often applied in microscale models. Loading conditions with different mode mix ratios can be achieved by rotating the FE models with specific angles [9]. Since loads are generally applied at rates in the order of mm/min, the dynamic effects are negligible and most simulations are quasi-static.

To save the computation cost, macro-scale finite element models have to be developed to simulate the mechanical actions of full-scale structures containing multiple Z-pins. It has been proved in the micro-scale models that individual Z-pins behave according to non-linear traction–separation laws during the reinforcement and failure process. As a result, Z-pins can be reasonably approximated by nonlinear elements or forces obedient to traction–separation laws to reduce computation cost. The individual Z-pins in macro-scale models can be simulated as spring elements, cohesive elements, interface elements, beam elements, truss elements or forces applied at nodes. The benefits and limitations of each simulation method are summarized in Table 4.1.

**Table 4.1** Comparison of diverse simulation methods of Z-pin reinforcements

| Simulation method | Benefits | Limitations |
|---|---|---|
| Spring elements | • Simulate each Z-pin individually<br>• Exert concentrated traction forces<br>• Employ user-defined force–displacement relation through tabular inputs | • Solely simulate axial stress components<br>• Require knowledge of exact locations of Z-pins<br>• Show mesh dependence |
| Cohesive elements | • Facilitate the simulation of both discrete and continuous toughening mechanisms<br>• Facilitate the simulation of Z-pin responses under complex loading conditions<br>• Show mesh independence | • Necessitate meshing refinement at the Z-pin locations<br>• Require recalculation of properties or re-meshing with changed Z-pinning configuration<br>• Employ simplified bilinear or trilinear traction–separation laws |
| Truss elements | • Efficiently simulate structures with large numbers of Z-pins<br>• Display low computational cost, high flexibility and high mesh independence | • Require very fine structural-level meshes<br>• Fail to simulate Z-pins under mode II or mixed-mode loading |
| Interface elements | • Able to employ general nonlinear force–deflection relationships | • Show mesh dependence |
| Beam elements | • Account for rotation and transverse shear deformation | – |
| Nodal forces | • Achieve high computational efficiency<br>• Simulate each Z-pin individually<br>• Replicate the physical reality | • Show mesh dependence |

One of the most common strategies is to simulate the Z-pin bridging actions by spring elements. Springs connect pairs of nodes on the prospective delamination surfaces. This method has been used in the simulation of Z-pinned CFRP composites subject to mode I/II or mixed-mode loading and Z-pinned joints under shear, bending, or pull-off loading. However, spring elements representing Z-pins must be inserted at nodes of existing meshing, which potentially leads to overly fine or inefficient meshes. In addition, only the axial stress components can be simulated by 2D spring elements. The radial components are ignored and mode I/II bridging forces are forced to follow the same pattern. The reinforcements of Z-pins can often be modeled using cohesive elements. This approach facilitates the simulation of large structures, to which it is hard to apply spring elements.

There are two typical methods to simulate the reinforcements of numerous Z-pins in composite laminates: (1) homogenizing the bridging force of Z-pins and representing it by an equivalent reinforced interfacial property; (2) embedding discrete nonlinear elements at the exact location of each Z-pin. The Z-pin bridging force can be smeared over the whole reinforced area in the equivalently distributed models. The Z-pin locations are reflected by the smeared properties via 'bridging maps', so explicit geology of Z-pins is unnecessary. In addition, this approach facilitates the simulation of random Z-pin arrangements or a large number of Z-pins. On the other hand, such modelling only yields high accuracy for the laminates reinforced by Z-pins with small diameters and close spacing [10]. Moreover, these models are exclusively valid for predefined loading conditions, as the smeared properties are determined according to mode mix ratios and have to be calibrated in advance [11]. When Z-pin parameters or loading conditions are modified, it is necessary to recalculate the smeared properties. The latter discrete approach is more efficient when parametric optimization and failure mechanics analysis are involved. This approach avoids using average or concentrated Z-pin bridging force and contributes to a discrete reinforcement rather than a continuous mechanism. This discrete representation method of Z-pins advances the study of the unstable and dynamic nature of delamination growth. However, the exact position of each Z-pin must be pre-known and fine meshing at Z-pin locations is required to achieve convergence. Changes in Z-pinning parameters may necessitate the re-meshing of structure.

Traction–separation laws (also known as pullout models) could accurately characterize the relationship between traction and pullout separation and reflect the bridging mechanisms of Z-pins in laminates, linking the micro-scale and macro-scale models together [7]. These traction–separation laws can be derived from pullout force–displacement results of the forementioned micro-scale modelling, experiments or theoretic analysis.

Cohesive Zone Model method has been successfully applied to simulate the predefined crack propagation, based on the assumption that failures occur over a strip zone ahead of the crack tip. It facilitates the modelling of continuum delamination, overcoming numerical difficulties and showing great simplicity. It also enables a unified representation of whole crack evolution within one model [12], in which the initiation and propagation of crack can be predicted by implementing (quasi-) zero-thickness cohesive elements along pre-defined crack paths [12]. This kind of pre-defined crack path along the mid-plane also can be modeled by two sub-beams with nods coupled with an offset position at the interface [13]. A decohesion element method has been developed to predict both onset and propagation of delamination without predefined crack paths [14].

Delamination of composite can be determined by fracture energy. The delamination will advance if the energy release rate of the specimen is greater than the critical intrinsic toughness. The Strain Energy Release Rate (SERR) at the crack tips in the DCB, ENF and impact specimens can be calculated by Virtual Crack Closure Technique (VCCT). This approach shows computational efficiency since energy release rates in diverse directions can be calculated in a single analysis. The J-integral and contour integral methods also can be applied. The energy dissipation due to Z-pins is calculated by comparing the energy release rates of one contour involving Z-pins and another contour excluding Z-pins [15].

In addition to the criteria used in micro-scale models, the critical crack opening displacement (COD) criterion, critical stress criterion and maximum nominal stress criterion have also been used in macro-scale models to determine the crack initiation, crack growth and final failure of DCB, ENF and MMB specimens.

The sub-beams of specimens can be modeled as beam elements, plane strain elements, shell elements or solid elements. To accurately capture the rotations, 8-node selectively reduced solid elements with hourglass control can be used to model the laminate beams. The interfaces between two sub-beams are often modeled by embedding a layer of cohesive elements along the bonded region. The pre-existing cracks in test specimens are often modeled using repetitive and disconnected nodes.

Loading apparatus can be simulated by rigid elements or solid elements. Reinforcement tabs are usually meshed using 3D solid elements. Displacement-controlled loading conditions are always applied, in which separation/transverse/compressive displacements are imposed according to the loading conditions. Loading points or tabs are connected to the free surfaces of laminates through rigid body constraints or multi-point (i.e., master node) coupling constraints whereby all degrees of freedom are coupled with slave surfaces. In the case of mode I/II loading, the upper node or loading point is always constrained in the x- and y-directions, and the bottom node or two lower rollers are usually fixed in all three axes/translational directions.

### *4.1.2 Multiscale Simulation Framework*

Composites are typically multiscale and each constituent influences the mechanical behaviours of macroscopic materials. By including detailed descriptions of microscale characteristics, high-fidelity microscale modelling can be realised at the expense of computation time. To achieve high computational efficiency, some microscale details must be ignored when performing macroscale modelling. By bridging microscale and macroscale analysis through information transmission, multiscale simulation can achieve a suitable compromise between efficiency and accuracy.

Although separate microscale and macroscale simulations of Z-pinned composites have been conducted, there have been only a few multiscale FE simulations of Z-pinned composites laminates. Bianchi and Zhang [16] analysed the mode-I delamination properties of thin Z-pinned composites laminates. In their study, only Z-pin pullout was simulated as a major failure mechanism and Z-pin rupture was ignored. The resultant cohesive law was applied to represent the enhanced delamination properties resulting from Z-pins. Bianchi et al. [17] also applied a multiscale modelling approach to study the structural behaviours of Z-pinned T-joints. The bridging force of each Z-pin under tensile and shear stress conditions was calculated using a unit cell model. The resulting predictions were transformed into Z-pin traction–separation laws and then used in a macroscale model. Under pure mode-I loading, Z-pin reinforcement was completely attributed to friction, and the interfacial debonding process was completely ignored. The shear deformation of Z-pins under mixed mode-I/II loading was analysed as a bar braced by an elastic laminate foundation. The associated macroscale model studied the structural-level mechanics through the application of the CZM method. The cracks along the centre of the web, skin/flange interface, and central line of the stiffener were simulated by inserting cohesive elements. Every composite sub-beam was modelled as a continuum shell layer. Such multiscale analyses of Z-pinned composites typically adopt a hierarchical bottom-up strategy, where Z-pin bridging information is passed from the microscale to the macroscale level. Microscale analyses represent the bridging and failure behaviours of the individual Z-pins in detail and macroscale studies predict the macroscopic mechanical properties of composite reinforced by numerous Z-pins subjected to various loading conditions.

CZMs have been broadly applied to simulate microscale damage and macroscale delamination cracks in Z-pinned composites to overcome numerical difficulties and achieve simplicity. A thin or zero-thickness layer of cohesive elements is inserted at predefined dominant damage locations. In contrast to the case of sudden failure in classic fracture mechanics, CZMs facilitate the modelling of continuum failure until complete failure occurs. Additionally, CZMs comprehensively reflect the entire failure procedure from crack initiation, and progressive degradation to final failure. The traction generated by CZMs is determined by the displacement between integral points based on traction–separation laws. Recently, CZMs have been employed to simulate the interfacial debonding, rupture, and splitting of Z-pins in detail, as well as the macroscopic interlaminar delamination of Z-pinned composites specimens under mode-I, mode-II, and mixed-mode loading. Predefined damage can also be modelled as decohesion elements or two sub-beams with coupled nodes. Additional details regarding the multiscale simulation methodology and damage growth simulation applied to Z-pinned composites laminates can be found in our previous review [4].

We establish a multiscale modelling framework to determine the effects of Z-pin/composite interfacial cracks on the microscopic bridging mechanisms and macroscopic mode-I interlaminar fracture properties of Z-pinned composites (Fig. 4.1). Microscale models simulate the microstructures and micromechanisms of Z-pinned unit cells and predict the effects of diverse Z-pin/composite interfacial conditions on Z-pin pullout responses. The obtained Z-pin bridging information is passed to

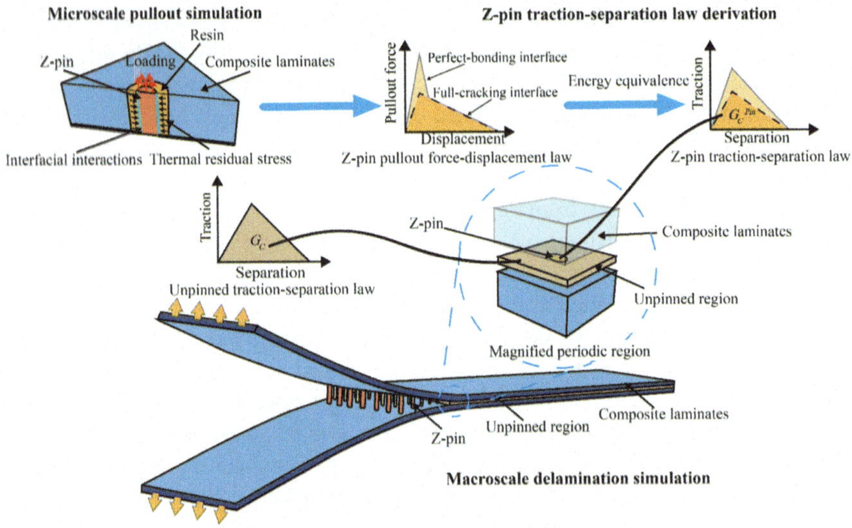


**Fig. 4.1** Multiscale simulation strategy for Z-pinned composites

macroscale simulations through Z-pin traction–separation laws. Macroscale models are used to predict the bridging behaviours of numerous Z-pins and the delamination properties of Z-pinned composites. The setup of macroscopic models adopted in this study is based on the experiments published in Ref. [18] and the microscale models represent the minimal periodic units that are extracted.

(1) Microscale pullout simulation

Unit cell models are established to study the microscopic effects of interfacial cracks between Z-pins and surrounding composite. Perfect-bonding and full-cracking Z-pin/composite interfaces are considered two opposing scenarios. Each unit cell model involves a single Z-pin, the surrounding resin, and unaffected composite. Microscale interactions such as elastic stretching, interfacial debonding, and frictional sliding are modelled. Radial thermal residual stress is simulated as compressive pressure. The obtained pullout force–displacement relationships of individual Z-pins reflect the effects of Z-pin/composite cracks on microscale Z-pin bridging mechanisms.

(2) Z-pin traction-separation law derivation

Z-pin traction–separation information passes from the microscale pullout models to the macroscale delamination models. Including the effects of Z-pin/composite interfacial conditions, the microscale pullout force–displacement results are transformed into traction–separation laws based on the equivalence of energy dissipated by Z-pins to achieve computational cost savings on the macroscale level.

(3) Macroscale delamination simulation

Macroscale models are used to investigate the effects of Z-pin/composite interfacial cracks on the mode-I delamination properties of Z-pinned composites. The mechanical responses of Z-pins and unpinned regions follow separate cohesive traction–separation laws. The delamination properties of Z-pinned composites with diverse Z-pin/composite interfacial conditions are predicted using macroscale models and further validated based on experimental results.

### *4.1.3 Z-pin/Composite Interfacial Conditions*

Figure 4.2a shows how Z-pinning results in wavy fibres and resin-rich regions. Z-pins push the fibres aside and eyelet-shaped cavities are generated. Before gelation and solidification, the resin flows and fills these cavities, creating resin-rich regions. The Z-pins and surrounding resin-rich regions experience different volume changes during the curing process based on the discrepancy between their thermal expansion coefficients. This mismatch results in high interfacial stress around the Z-pins, which is likely to cause the tiny Z-pin/composite interfacial cracks shown in Fig. 4.2b [19]. These cracking defects significantly alter the performance of Z-pinned composites.

The microscale effects of curing-induced Z-pin/composite interfacial cracks are first investigated in unit cell models. Two cases with perfect-bonding or full-cracking Z-pin/composite interfaces are simulated separately. A Z-pin/composite interface can be assumed to be perfectly bonded if the curing-induced interfacial stress is less than the interfacial strength. In this case, a Z-pin is initially bonded to composite through

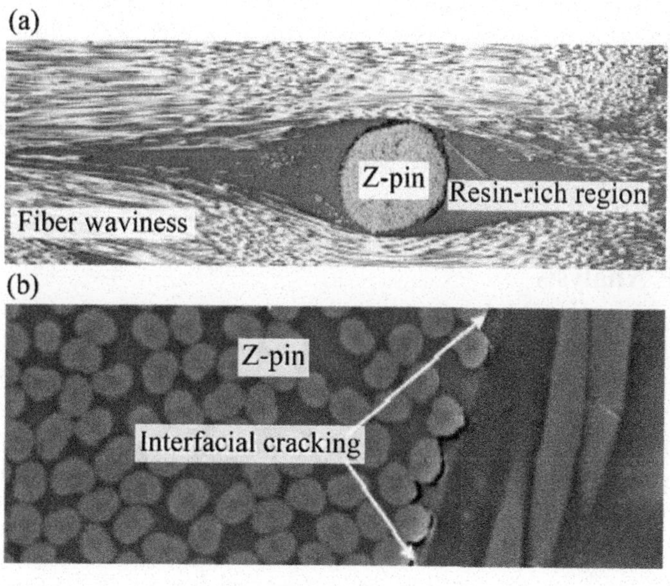


**Fig. 4.2** Typical microstructures of Z-pinned composites [19]

chemical bonding. However, under increased loading, the Z-pin experiences elastic stretching and interfacial debonding prior to frictional sliding. The stress increases linearly until it reaches the critical strength $T_0$ at the corresponding displacement $\delta_0$. Debonding occurs as the stiffness degrades and fails at the maximum displacement $\delta_{\max}$.

In the other scenario, the interface between the Z-pin and adjacent laminates is assumed to be fully debonded after curing. In this case, frictional sliding is the dominant bridging mechanism. The friction along the Z-pin boundary can be simulated as the sum of the standard Coulomb friction and residual friction component as follows [20]:

$$\tau = \max(\tau_{\max}, \tau_0 + \mu\sigma_n), \tag{4.1}$$

where $\tau$ represents the frictional stress during sliding; $\tau_{\max}$ is the maximum allowable stress, $\tau_0$ is the residual friction force, $\sigma_n$ is the normal stress caused by compressive pressure, and $\mu$ is the Coulomb friction coefficient. The frictional force increases gradually until it reaches the maximum allowable stress and then remains constant.

Because the thermal expansion coefficient of a Z-pin is typically one order of magnitude less than that of base materials, Z-pins shrink significantly less during the cooling process. Therefore, residual stress compresses Z-pins along the radial direction. This radial thermal residual stress can be estimated as follows [21]:

$$\sigma = \frac{\Delta T(\alpha_{resin} - \alpha_{z-pin})E_{resin}E_{z-pin}}{E_{resin} + E_{z-pin}}, \tag{4.2}$$

where $\Delta T$ is the temperature reduction during the cooling process, and $\alpha_{z-pin}$ and $\alpha_{resin}$ represent the thermal expansion coefficients of the Z-pin and surrounding resin, respectively. $E_{z-pin}$ and $E_{resin}$ are Young's moduli of the Z-pin and resin, respectively. This radial compressive pressure is applied along the Z-pin boundary as a predefined load. The non-uniform distribution of residual stress that appears in practice is not included here and will be considered in detail in our future work.

## 4.2 FE Analysis

In this section, FE analysis is conducted to implement our three-step multiscale simulation strategy. First, the microscale FE model represents the effects of Z-pin/composite interfacial cracks on the bridging mechanisms of a single Z-pin (Section 4.2.1). The Z-pin bridging information flows from microscale to macroscale models through traction–separation laws, as discussed in Section 4.2.2. The macroscale FE model then predicts how Z-pin/composite interfacial cracks influence the delamination responses of Z-pinned composites.

### 4.2.1 Microscale FE Model

Unit cell models simulate the representative microstructures and micromechanisms of Z-pinned composites. The results of experimental observations indicate that Z-pin pullout is the dominant failure pattern of Z-pinned composites under mode-I loading. Therefore, only Z-pin pullout failure is considered in our model. Because microscale changes with low-density Z-pins are confined to a small region, the microscale model is established with length and width values equal to the spacing of the Z-pins in the macroscale model (3.1 mm in this case). According to the periodicity of Z-pinned specimens, each unit cell model involves a Z-pin, surrounding resin, and unaffected laminates. Under tensile loading, individual Z-pins are typically extracted from one composite sub-beam and remain embedded in another. Therefore, the unit cell models simulate only one sub-beam and the total displacement is equal to half of the specimen thickness. More precisely, unit cell models are defined with 12 piles of IMS/924 unidirectional prepregs with a nominal thickness of 1.5 mm and carbon fibres are aligned along the specimen length. Z-pins with a radius of 0.255 mm are defined as unidirectional T300 carbon fibre tows saturated with bismaleimide (BMI) resin. As reported in Refs. [22], the eyelet-shaped resin region is simplified as a cylindrical wall. The external radius of the resin-rich region is 0.51 mm. Eight-node linear brick elements with reduced integration and hourglass control (C3D8R elements in ABAQUS) are employed to simulate the Z-pin and surrounding resin regions, and eight-node continuum shell elements with reduced integration and hourglass control (SC8R elements) are used to simulate the bulk composite. The application of shell elements that simulate thin composite laminate structures can achieve reasonable accuracy and reduce computation time. The layup of composite laminates is defined through the 'composite layup control' in the 'property module' and one layer of mesh is applied through the thickness. Relevant material properties can be found in Table 4.2 [16]. Both the composite and Z-pins are defined as orthotropic materials with linear elasticity. The pure resin is considered to be isotropic and homogeneous.

Cohesive behaviour is used to model the progressive debonding process of the Z-pin/composite interfaces by defining the interfacial fracture toughness ($G_C$),

**Table 4.2** Mechanical properties of composite, Z-pins, and resin [16]

| Composites (IMS/924) | | | | | | | | |
|---|---|---|---|---|---|---|---|---|
| $E_1$(MPa) | $E_2$(MPa) | $E_3$(MPa) | $G_{12}$(MPa) | $G_{13}$(MPa) | $G_{23}$(MPa) | $\nu_{12}$ | $\nu_{13}$ | $\nu_{23}$ |
| 138,000 | 11,000 | 11,000 | 4400 | 4400 | 3920 | 0.34 | 0.34 | 0.4 |
| Z-pin (T300/BMI) | | | | | | | | |
| $E_1$(MPa) | $E_2$(MPa) | $E_3$(MPa) | $G_{12}$(MPa) | $G_{13}$(MPa) | $G_{23}$(MPa) | $\nu_{12}$ | $\nu_{13}$ | $\nu_{23}$ |
| 100,000 | 30,000 | 30,000 | 4600 | 4600 | 3900 | 0.35 | 0.35 | 0.4 |
| Resin | | | | | | | | |
| $E$(MPa) | | | | $\nu$ | | | | |
| 11,000 | | | | 0.4 | | | | |

**Table 4.3** Modelling parameters of unit cell models [24]

| Debonding parameters | | | Friction parameters | | | Curing parameters | | |
|---|---|---|---|---|---|---|---|---|
| $K$ (N/mm$^3$) | $T_0$ (MPa) | $G_c$ (N/mm) | $\mu$ | $\tau_0$ (MPa) | $\tau_{max}$ (MPa) | $\Delta T$ (°C) | $\alpha_{resin}$ (k$^{-1}$) | $\alpha_{Z\text{-pin}}$ (k$^{-1}$) |
| $1 \times 10^5$ | 36 | 2.62 | 0.25 | 1.25 | 40 | 155 | $2.4 \times 10^{-5}$ | 0 |

$K$ cohesive stiffness; $T_0$ maximum strength; $G_c$ interfacial fracture toughness; $\mu$ Coulomb friction coefficient; $\tau_0$ residual friction force; $\tau_{max}$ maximum allowed shear stress; $\Delta T$ temperature reduction; $\alpha_{resin}$ thermal expansion coefficient of resin; $\alpha_{Z\text{-pin}}$ thermal expansion coefficient of Z-pin

maximum strength ($T_0$), and cohesive stiffness ($K$). Z-pin/composite interfacial debonding along different directions is assumed to possess consistent properties. The interfacial fracture toughness and maximum strength are calculated based on experimental results [23], reflecting the combined impact of mode-I and mode-II fractures. The cohesive stiffness is set to achieve convergence. Debonding initiation is predicted by the quadratic nominal stress criterion and the subsequent stiffness degradation of bonding cohesive elements is determined by a power law with a power of one. Once interfacial bonding begins to degrade, the Z-pin surface undergoes modified Coulomb friction (Eq. 4.1). The temperature decreases from 180 to 25 °C during post-cure cooling, so the resulting radial thermal residual stress can be calculated based on Eq. (4.2). Relevant modelling parameter values are listed in Table 4.3 [24].

The external surface of the resin region is connected to adjacent composite using 'tie' constraints. Mesh sensitivity analysis is conducted: 0.05 × 0.05 mm meshing is employed in the surrounding composite and 0.025 × 0.025 mm meshing is employed in the Z-pin and resin components. Displacement-controlled tensile loading is applied to the Z-pins while the bottoms of the composite and resin regions were constrained.

### 4.2.2 *Z-pin Traction–Separation Law*

The pullout force–displacement relationships of individual Z-pins obtained from the unit cell models described above are transformed into cohesive traction–separation laws that further represent the bridging behaviours of numerous Z-pins in macroscopic composite structures. The traction $T$ of each Z-pin can be calculated according to the following equation [21] based on the microscale prediction of $F(u)$:

$$T(u) = \frac{F(u)}{\pi r^2}, \tag{4.3}$$

where $F(u)$ is the predicted pullout force of individual Z-pins at the corresponding displacement $u$, and $r$ represents the Z-pin radius.

Based on fracture energy equivalence, this traction–separation law is simplified to a bilinear form to simplify setup and reduce computational cost. The fracture energy of a Z-pin $G^{pin}$ can be calculated using the following formula [25]:

$$G^{pin} = \int_{0}^{u} T(u)\mathrm{d}u. \tag{4.4}$$

In other words, the area under the traction–separation curve is equal to its bilinear counterpart. This equivalence reflects the essence of the Z-pin bridging mechanism, which absorbs energy during failure. The numerical accuracy of this transformation is verified by Bianchi and Zhang [16] and a similar strategy has been employed in a previous study [21]. The influence of different interfacial strengths under this simplification is further discussed in the parametric section.

### 4.2.3 Macroscale FE Model

Each macroscopic specimen is comprised of 24 piles of IMS/924 unidirectional prepregs with a nominal thickness of 3 mm. A thin polyamide film is inserted along the centre plane of the specimen, forming an initial crack of length 60 mm. The Z-pinned region is 25 mm in length and spanned the entire width of the specimen. T300/BMI Z-pins are inserted with a spacing of 3.1 mm, which corresponds to the setup of the unit cell models.

The bulk composites are simulated using a continuum shell layer (SC8R elements). Each element in the thickness direction contains all plies defined in the ply table. The initial crack region is simulated as an unbonded area between the two sub-beams. Cohesive elements (COH3D8 elements) with a thickness of 0.1 mm are applied to simulate both Z-pinned and unpinned regions along the predefined crack path. The Z-pinned and unpinned regions follow two separate cohesive laws. Z-pin cohesive laws with diverse Z-pin/composite interfacial conditions are calculated based on microscale predictions and the corresponding parameter values are presented in Table 4.4. The cohesive parameters for unpinned regions are presented in Table 4.5 and are validated through unpinned double-cantilever beam simulations. The quad damage criterion is used to determine the crack initiation of the cohesive elements and a linear energy criterion is used to identify final failure. When the failure criterion is satisfied, the corresponding cohesive elements are deleted. Damage energy is dissipated irreversibly during the Z-pin failure and crack propagation process. Our analysis only considers dominant delamination along the central plane caused by the predefined initial crack. In practice, a delamination crack propagates through an unknown pathway or branches through bulk composite because Z-pins introduce microscale damage and fibre misalignment. Simulation methods, including the phase-field approach, augmented FE method, and extended FE method, can be used to model the complex crack branching and curved paths of Z-pinned composites [26].

**Table 4.4** Derived modelling parameters for Z-pin cohesive elements in macroscale models

| Perfect-bonding Z-pin/composite interface | | | Full-cracking Z-pin/composite interface | | |
|---|---|---|---|---|---|
| $K$ (N/mm$^3$) | $T_0$ (MPa) | $G_c^{pin}$ (N/mm) | $K$ (N/mm$^3$) | $T_0$ (MPa) | $G_c^{pin}$ (N/mm) |
| 2956 | 148 | 111 | 2317 | 116 | 87 |

$K$ cohesive stiffness; $T_0$ maximum strength; $G_C^{pin}$ fracture toughness of Z-pins

**Table 4.5** Modelling parameters for unpinned cohesive elements in macroscale models

| $K$ (N/mm$^3$) | $T_0$ (MPa) | $G_c^{unpin}$ (N/mm) |
|---|---|---|
| $1 \times 10^5$ | 30 | 0.3 |

$K$ cohesive stiffness; $T_0$ maximum strength; $G_C^{unpinned}$ fracture toughness of unpinned regions

Mode-I loading is applied under displacement control, meaning opening displacements were implemented at the sub-beam ends. Based on the mesh size sensitivity, a 0.5 × 0.5 mm mesh is applied to free sub-beams, and 0.5 × 0.1 mm and 0.5 × 0.25 mm meshes are applied to the Z-pinned and unpinned regions, respectively. A total of 459,097 elements were generated in our model.

## 4.3 Model Validation

This section presents the validation of the proposed multiscale simulation strategy based on experimental measurements with a focus on the bridging and failure behaviours of Z-pinned composites. The multiscale influence of Z-pin/composite interfacial cracks is also discussed.

The microscale model represents the progressive pullout of a single Z-pin and the evolution of the bridging force (Fig. 4.3a). Stress is constrained to the vicinity of the Z-pin and bulk composite are only marginally affected. The purple line represents the pullout response of a single Z-pin with perfect interfacial bonding and the brown line corresponds to the response with complete interfacial cracking. These two bridging laws are transformed into bilinear traction–separation laws, which are used in macroscale models to represent the bridging behaviours of numerous Z-pins.

With perfect interfacial bonding to the surrounding resin, the Z-pin exhibits a three-stage bridging response. Initially, the Z-pin deforms elastically and the bridging force increases monotonically. The combination of adhesion and static friction contributes to a sharp slope and large bridging limit, and the elastic stretching of the Z-pin and surrounding composite results in initial displacement. As the displacement grows, the load eventually reaches the interfacial strength. The interfacial debonding between the Z-pin and surrounding resin initiates and progresses, and is characterised by an abrupt load drop. The partial adhesion and sliding friction generate bridging traction during this stage and the Z-pin is then extracted under the sliding

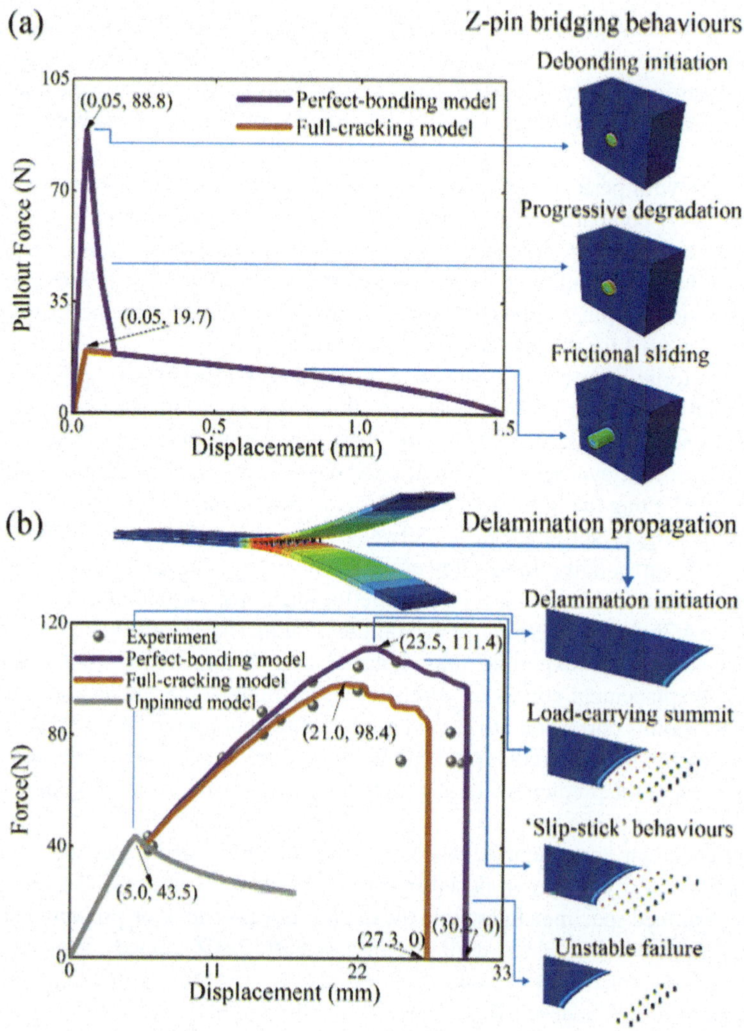


**Fig. 4.3** FE models and simulation results. **a** The bridging behaviors of a single Z-pin under the pullout force; **b** The delamination properties of Z-pinned composites under mode-I loading

frictional force. This three-stage failure pattern has been verified through experiments and numerical analyses [27]. However, the debonding and frictional stages tend to coalesce into a wide peak under large mode mixing angles, leading to less energy dissipation based on pin rupture and reduced frictional pullout.

If the Z-pin has completely separated from the bulk composite at the end of the curing process, then mechanical debonding and the resultant load drop will be absent, as indicated by the brown line. This behaviour has also been observed in the literature [8]. Interfacial friction governs the bridging response after the slight elastic stretching of the Z-pin in the full-cracking case. The contact area between the Z-pin and surrounding resin decreases as pullout proceeds and the associated bridging force decreases progressively. The frictional force suddenly drops to zero in the final stage, reflecting smaller friction resistance based on reduced contact pressure near the free surface.

A Z-pin with perfect interfacial bonding can carry a greater maximum traction load (88.8 N) compared to its counterpart with a full-cracking interface (19.7 N). Reflected by the areas beneath the pullout force–displacement curves, the energy absorbed during the pullout of each Z-pin decreases by more than 21% as a result of curing-induced Z-pin/composite interfacial cracks (22.6 mJ for the perfect-bonding case and 17.8 mJ for the full-cracking case).

The periodic insertion of Z-pins with a low density enables the application of a representative volume element (RVE) in this case. In composite specimens with closely spaced Z-pins, the interactions between Z-pins become significant. The nearby resin-rich eyelets overlap and coalesce into a ceaseless resin channel that is orientated along the fibre direction. This results in localised low stiffness and can potentially trigger splitting cracks between fibres. The fibre alignment and stress distribution surrounding the Z-pins are also changed. A larger RVE with periodic boundary conditions must be applied to define such large-scale interactions.

Figure 4.3b depicts the progressive failure of numerous Z-pins under the opening displacements. The experimentally measured and numerically predicted load-opening displacement curves of unpinned and Z-pinned specimens are compared. The black points correspond to the experimental results from Ref. [18] and the grey line represents the FE prediction of an unpinned specimen. The purple and brown lines represent the predictions of Z-pinned specimens with diverse Z-pin/composite interfacial conditions.

The excellent agreement between the FE predictions and experimental results proves the validity of the proposed multiscale simulation strategy. The test results for the Z-pinned specimen fall within the bounds of FE simulation predictions. These FE models also capture the overall response evolution of Z-pinned specimens under mode-I loading, providing accurate descriptions of resin failure, Z-pin reinforcement, and final failure. Under tensile displacements, the force increases linearly initially. The Z-pin bridging zone has not been completely developed and delamination initiation relies on the matrix toughness. Therefore, the delamination initiates at the same point (about 43.5 N) for both Z-pinned and unpinned samples. This ineffectiveness of Z-pins for halting initial cracking has also been observed experimentally [28]. Additionally, Z-pins have been found to have little effect on the stiffness of composite specimens, so the initial slopes of the Z-pinned and unpinned specimens are the same. Following a localised peak, the load of the unpinned specimen decreases monotonically as the catastrophic delamination grows along the predefined path. In contrast, the Z-pinned specimen reloads when the crack travels through the first row of Z-pins. Z-pin rows bridge the crack wake, and the delamination tip is sheltered from the opening force. Therefore, the delamination growth decelerates and the specimen can sustain an increased load. With further displacement, an increasing number of Z-pins become involved until they become saturated at the force summit. The maximum load-carrying capacity of the Z-pinned specimens (98.4–111.4 N) is more than two times greater than that of the unpinned case (approximately 43 N). The curves then begin to oscillate based on the 'slip-stick' behaviours of the Z-pins [29]. The deactivation of a pin row at the far end of the bridging area accompanies the activation of the pin row along the delamination wake. The number of active Z-pins remains approximately constant during this period. This oscillating phenomenon can be alleviated when the distributed reinforcement of Z-pins is achieved. Once the crack tip passes beyond the reinforced zone, the number of active Z-pins is reduced and unstable failure occurs with decreasing crack resistance, as observed experimentally [18].

The simulation results also demonstrate the effects of curing-induced Z-pin/composite interfacial cracks on the macroscopic mechanical responses of Z-pinned composites. The two FE predictions begin bifurcating after resin failure. The curing-induced interfacial cracking contributes to a reduction in maximum load-carrying capacity. The corresponding failure displacement also slightly decreases as a result of the weaker reinforcement of Z-pins with initial interfacial cracking. The case with perfect bonding exhibits a longer, but less pronounced fluctuation following the load peak, indicating that the perfect-bonding interfaces between Z-pins and composite result in an extended Z-pin bridging zone and reduce the failure intensity of Z-pins.

Figure 4.4 presents the peel stress of Z-pinned composites laminates with different Z-pin/composite interfacial conditions under the same tensile displacement. The active Z-pins along the crack tip experience high peel stress and hinder catastrophic delamination propagation. The concave crack propagation with the maximum growth rate at the centre has also been successfully predicted [16]. A large-scale Z-pin bridging zone has been established, as indicated by several Z-pin rows becoming activated during the simulation. The simulation results also reflect the non-uniform distribution of the peel stress along the Z-pin columns. Z-pins generate larger tensile traction with an increased displacement near the crack tip. The traction then decreases in the rear bridging zone as the contact area decreases when Z-pins are almost pulled out. This tendency exhibits excellent agreement with the results of the microscale models. Compared to the Z-pins with full-cracking Z-pin/composite interfaces, the perfect-bonding Z-pins can sustain greater maximum peel stress, resulting in later activation, degradation, and failure of Z-pins.

## 4.4 Parametric Study

In this section, our simulation strategy is applied to investigate the multiscale roles of interfacial bonding, radial thermal residual stress, and interfacial friction along Z-pins. The microscale traction force and energy dissipation during single Z-pin pullout are presented in Figs. 4.5, 4.6 and 4.7, and the macroscale bridging and failure behaviours of Z-pinned composites are presented in Fig. 4.7.

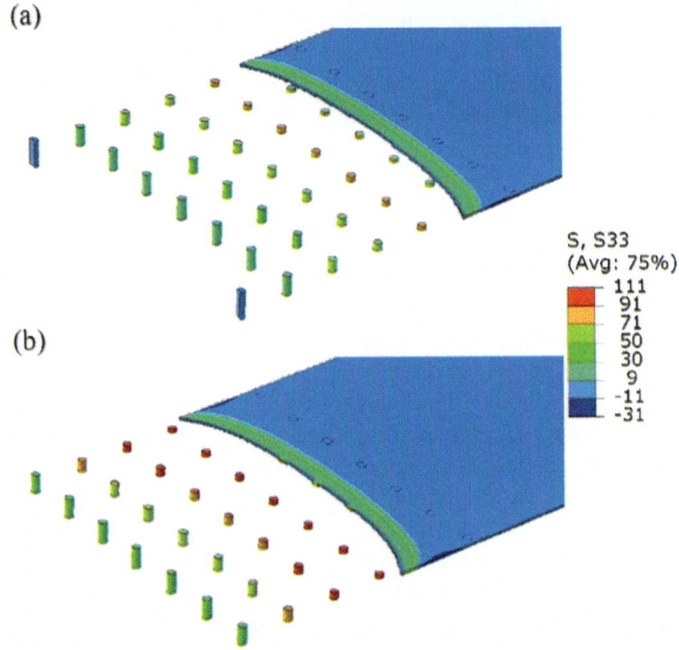


**Fig. 4.4** Peel stress at the delamination interfaces of Z-pinned composites laminates **a** with full-cracking Z-pin/composite interfaces and **b** perfect-bonding Z-pin/composite interfaces

## *4.4.1 Effects of Interfacial Bonding*

The Z-pin bridging response is affected by the Z-pin/composite interfacial bonding properties after the curing process. The bridging process will exhibit a smooth two-stage response if the Z-pin/composite interface is largely debonded. Otherwise, individual Z-pins will act according to a three-stage bridging law with a significant debonding load drop. The bridging mechanisms of Z-pins with full-cracking interfaces are dominated by friction and are unrelated to interfacial bonding properties. Therefore, we only consider perfect-bonding cases here.

The Z-pin/composite interfacial fracture toughness $G_C$ of 2.62 kJ/m$^2$ and maximum strength $T_0$ of 36 MPa is taken from Ref. [23]. Improved interfacial fracture toughness and maximum strength can be achieved by refining the Z-pinning technique or employing Z-pin surface treatments such as coating, cryogenic, or plasma treatments.

Figure 4.4 presents the effects of Z-pin/composite interfacial bonding parameters on microscale bridging mechanisms. The area under the pullout force–displacement curves represents the energy consumed by individual Z-pins. One common characteristic is that the Z-pin/composite interfacial bonding properties do not affect subsequent frictional sliding. It has also been observed that increased interfacial fracture toughness with a constant maximum strength significantly reinforces the debonding

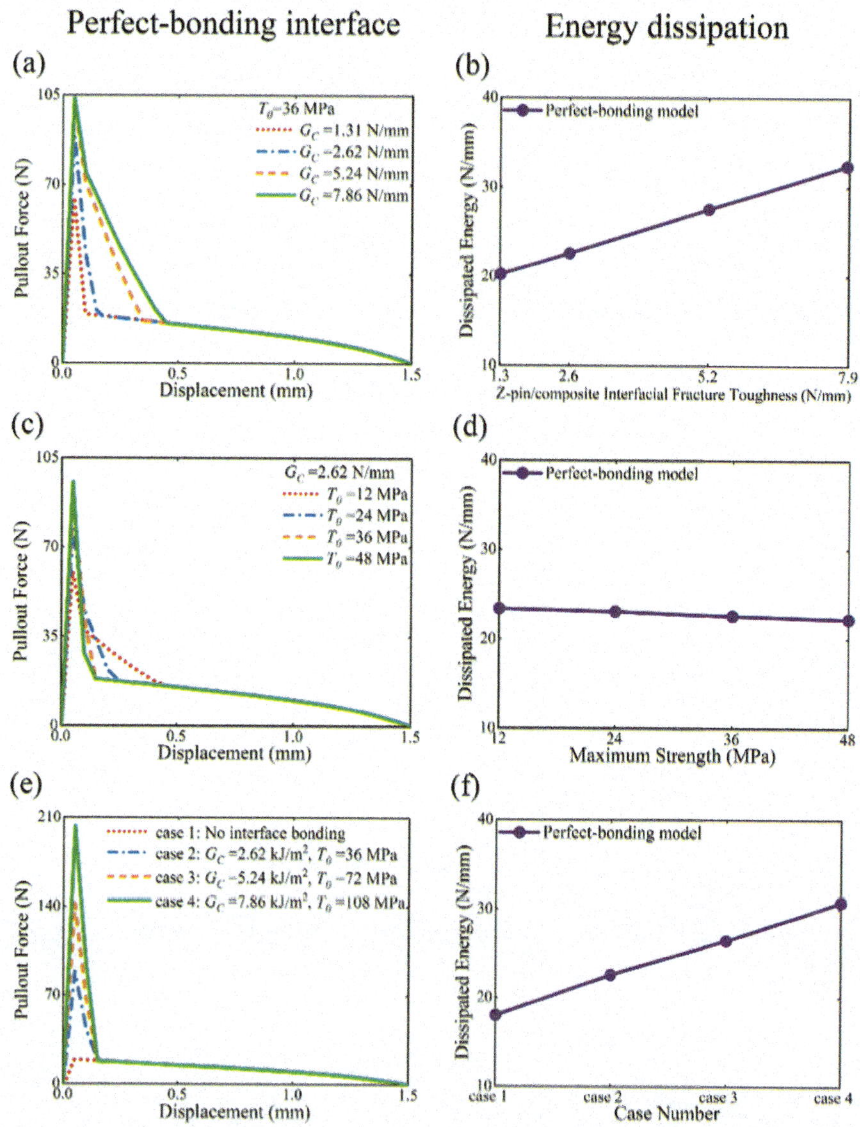


**Fig. 4.5** Effects of Z-pin/composite interfacial bonding parameters on the pullout force of a single Z-pin

process with a higher maximum pullout force, longer debonding stage, and greater energy dissipation (Fig. 4.5a, b). However, with the same interfacial fracture toughness, increasing the interfacial strength contributes to a sharper debonding process with slightly less dissipated energy (Fig. 4.5c, d). This can be attributed to postponed

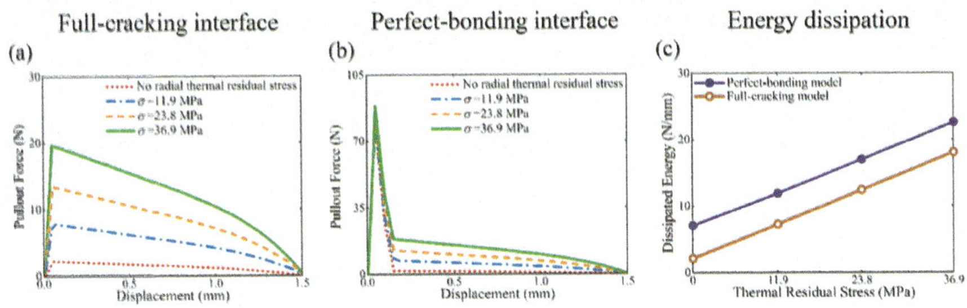


**Fig. 4.6** Effects of radial thermal residual stress on the pullout force of a single Z-pin

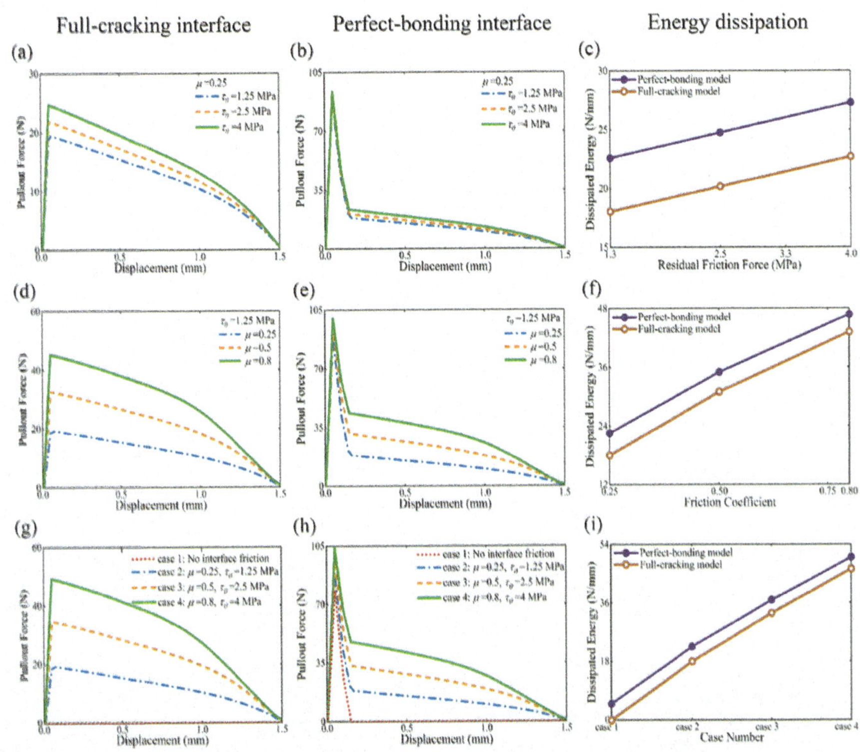


**Fig. 4.7** Effects of Z-pin/composite interfacial friction parameters on the pullout force of a single Z-pin

frictional sliding. Previous studies have verified the negligible influence of interfacial strength [30]. This also supports the simplified Z-pin traction–separation laws. Fracture toughness plays a more significant role than maximum strength in terms of determining Z-pin bridging behaviours. It should be noted that excessive interfacial fracture toughness will induce a failure transition from Z-pin pullout to rupture, even in a low-mode mixing region. If the fracture toughness and maximum strength of Z-pin/composite interfaces are changed synchronously, then perfect-bonding interfaces can sustain a higher maximum pullout force than full-cracking interfaces. Furthermore, the stronger the interfacial bonding properties, the greater the maximum pullout force and energy dissipation (Fig. 4.5e, f).

Figure 4.7a presents the macroscale effects of the fracture toughness and maximum strength of perfect-bonding Z-pin/composite interfaces. The full-cracking Z-pin/composite interfacial condition was simulated for comparison. The general bridging and failure trends are consistent, but the maximum load value increases with stronger interfacial bonding. The displacement of the final failure shifts to the right and the oscillation stage extends simultaneously. A steeper slope reflects the stronger reinforcement of Z-pins with an enhanced energy dissipation rate. The results also suggest that moderate Z-pin/composite interfacial properties can alleviate fluctuation during the slip-stick stage by forming a large bridging zone. With the further enhancement of interfacial bonding properties, the failure of each Z-pin becomes distinct, contributing to more significant fluctuations. From this perspective, the interfacial conditions between Z-pins and surrounding composite are essential for optimising the load-bearing capacity of Z-pinned composites specimens. In other words, the use of similar fibre and resin with bulk composite is a beneficial choice for Z-pins because it will yield excellent interfacial properties between Z-pins and surrounding composite.

### 4.4.2 *Effects of Radial Thermal Residual Stress*

A Z-pin/composite interface is subjected to radial compression after curing. This compressive thermal residual stress alters Z-pin performance as a result of contact pressure during Z-pin pullout. This section analyses the impact of the radial thermal residual stress on the bridging mechanisms and delamination properties of Z-pinned composites with diverse Z-pin/composite interfacial conditions.

Figure 4.7 presents the microscale effects of the radial thermal residual stress. In this discussion, the thermal residual stresses correspond to temperature changes of 155, 100, and 50 °C, and the case ignoring thermal residual stress serves as a control case. The bridging and failure patterns do not change once the Z-pin/composite interfacial conditions are determined. The maximum pullout force increases with a greater radial thermal residual stress as the additional normal stress promotes frictional clamping between the Z-pin and resin regions. As shown in Fig. 4.7c, the dissipated energy scales linearly with the radial thermal residual stress, regardless of interfacial conditions, and Z-pins with perfect-bonding interfaces exhibit higher reinforcement efficiency than their counterparts with complete debonding.

The effects of radial thermal residual stress on the mode-I interlaminar fracture responses of Z-pinned composites are presented in Fig. 4.8b and c. The maximum force and failure displacement increase with increasing radial thermal residual stress, regardless of the Z-pin/composite interfacial conditions. Compared to the curing-induced full-cracking cases, the specimens with perfect-bonding Z-pin/composite interfaces exhibit enhanced delamination resistance based on stronger interactions between Z-pins and composite. The perfect-bonding cases also exhibit longer oscillation.

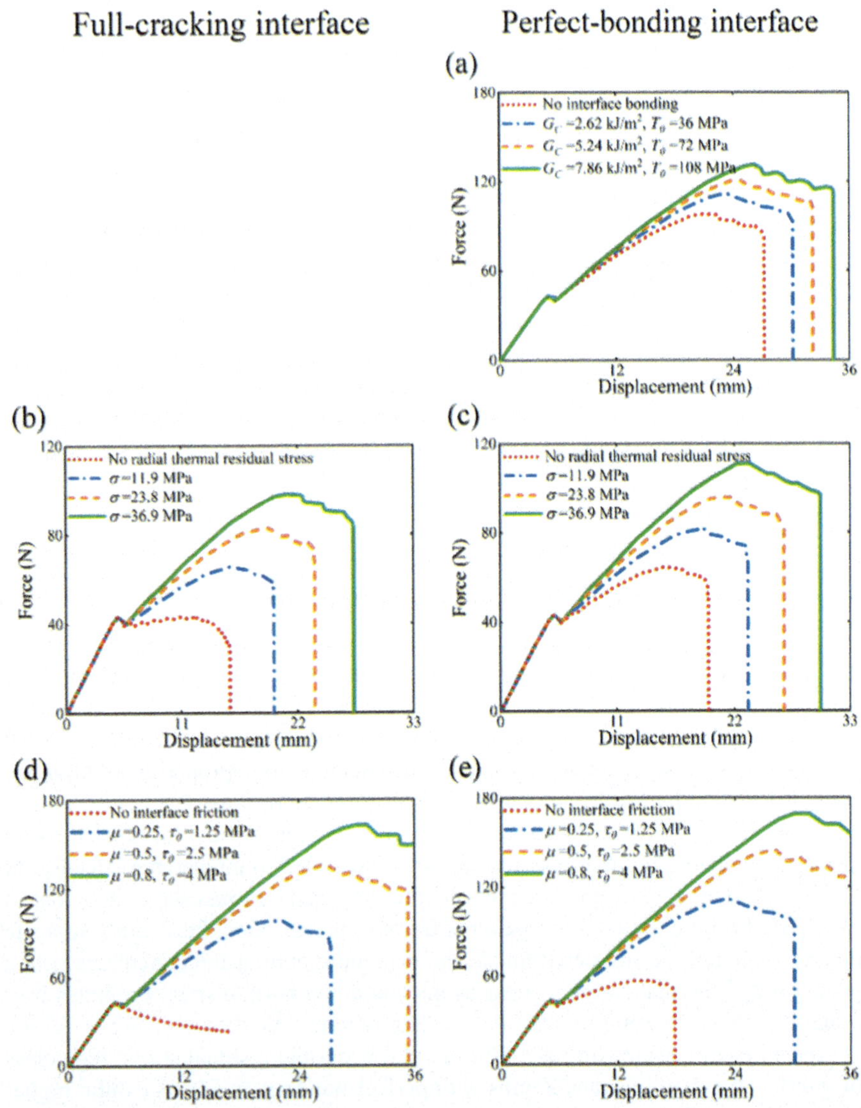


**Fig. 4.8** Effects of Z-pin/composite interfacial parameters on the force–displacement curves of Z-pinned composites

It should be noted that a significant temperature difference during the curing process does not necessarily guarantee superior Z-pin reinforcement. With a significant temperature reduction, the greater shrinkage mismatch between Z-pins and composite inevitably leads to residual stress around the Z-pins. Radial residual stress acts as normal compressive stress, improving the frictional force of individual Z-pins. However, axial stress stretches the composite, potentially inducing interfacial cracks along the Z-pins. Therefore, the cooling temperature drop should be optimised to balance radial and axial residual stress.

### *4.4.3 Effects of Interfacial Friction*

Residual friction and Coulomb friction govern the friction force during the sliding stage. This section discusses the effects of the residual friction force and Coulomb friction coefficient on the bridging mechanisms and delamination properties of Z-pinned composites with diverse Z-pin/composite interfacial conditions.

Figure 4.8 illustrates how friction parameters affect the pullout behaviours of a single Z-pin. The shape of the bridging laws is related to the initial interfacial conditions, friction coefficient, and residual friction force. With the same roughness, a Z-pin with a perfect-bonding interface can sustain a greater pullout force than its counterpart with complete debonding. The friction coefficient and residual friction force positively influence the energy dissipation during Z-pin failure, regardless of the Z-pin/composite interfacial conditions (Fig. 4.7c and f). An increased residual frictional force leads to an intensified peak force and steeper slope of the force–displacement response, which is in good agreement with Ref. [6]. Premature failure caused by an exaggerated Coulomb friction coefficient cannot be observed because Z-pin rupture is not considered here. Synchronous increases in the friction coefficient and residual friction force also result in a higher peak bridging force (Fig. 4.7g and h). Similar results have been reported elsewhere by comparing fibrous Z-pins to smoother metal Z-pins [31].

Figure 4.8d and e illustrate the dependence of the force–displacement relationship of Z-pinned composites under mode-I loading on interfacial friction parameters. Increasing the interfacial friction can enhance fracture toughness and delay final failure, as reported in Ref. [32]. Steeper curves reflect greater resistance to delamination. As a result, some specimens do not fail under the entire displacement loading range. With a synchronous increase in the friction coefficient from 0 to 0.25 and residual friction from 0 to 1.25 MPa, the maximum load doubles.

This study only considered the failure modes of interfacial debonding and frictional sliding because ideal mode-I loading was applied to thin composite plates. Z-pin rupture should be considered if the interfacial stress exceeds the rupture strength of the Z-pins. Rather than a gradual load reduction, the Z-pin rupture would cause an abrupt loss in the bridging load, as established in previous simulation analyses and experiments [33]. Thinner Z-pins inserted deeply into thicker composite (typically thicker than 8–10 mm) are more likely to rupture. With an increasing mixed-mode angle, the failure mechanism also transitions from Z-pin pullout to rupture. Another common failure pattern of mode-II specimens is snubbing. In this pattern, Z-pins are laterally pressed into the composite, generating local plastic deformation. Snubbing can increase the bridging force by three to ten times, resulting in greater energy dissipation. The frictional component would become more significant and stable, and the frictional bridging force would be mainly attributed to the side deflection of Z-pins, rather than thermal residual stress. Additionally, the Z-pins would also experience transverse shear fracture or internal splitting. The rupture, splitting, and snubbing of individual Z-pins can be simulated by inserting cohesive elements into each pair of adjacent elements. All of these microscale characteristics would introduce significant scatter into the mechanical responses of Z-pinned composites laminates.

### 4.4.4 Remarks

Our parametric analysis demonstrates that it is theoretically feasible to control the delamination behaviours of Z-pinned composites by customising the microscale bridging responses of individual Z-pins. Z-pin/composite interfacial conditions play a decisive role in microscale Z-pin bridging mechanisms. Z-pins with perfect-bonding interfaces exhibit a three-stage bridging mechanism, achieving increased energy dissipation, greater load-carrying capacity, and delayed failure of Z-pinned composites. With certain Z-pin/composite interfacial conditions, improved Z-pin reinforcement under pure mode-I loading conditions can be achieved by enhancing Z-pin/composite interfacial bonding, optimising the curing profile, or roughening the Z-pin surface. For example, if Z-pin/composite interfacial bonding properties are enhanced by selecting compatible material systems, delayed failure can be expected with a higher maximum bridging load and greater energy dissipation. Increasing the curing temperature difference to increase the radial thermal residual stress along Z-pins is another strategy for improving the bridging capabilities of Z-pins because compressive pressure influences Coulomb friction during the sliding process. However, it is important to compromise between Z-pin/composite interfacial cracking and compressive radial thermal stress along Z-pins. An alternative strategy is to roughen the surfaces of individual Z-pins to increase the residual frictional stress and friction coefficient, which can boost energy dissipation during the frictional pullout of Z-pins.

## 4.5 Conclusions

In this chapter, a multiscale framework is introduced to evaluate the mechanical properties of Z-pinned composites. Particularly, the delamination behaviours of Z-pinned composites considering the effects of Z-pin/composite interfacial cracks are investigated using the multiscale simulation framework. Unit cell models simulate the microstructures and micromechanisms of Z-pinned regions with diverse Z-pin/composite interfacial conditions. The derived Z-pin bridging information flow into macroscale models in the form of cohesive traction–separation laws. These macroscale models then predict the mechanical responses of composite specimens with numerous Z-pins, reflecting the macroscopic influence of Z-pin/composite interfacial cracks. The proposed multiscale strategy provides good accuracy in terms of predicting the bridging and failure responses of Z-pinned composites.

The simulation results indicate that curing-induced Z-pin/composite interfacial cracks may affect the interlaminar fracture responses of Z-pinned specimens by altering the microscale bridging mechanisms of individual Z-pins. Furthermore, parametric analysis reveals that enhanced interfacial bonding, radial thermal residual stress, and interfacial friction can reinforce the mode-I delamination properties of Z-pinned composites.

## References


1. Kumar, R.S., et al., Notch-strength prediction of ceramic matrix composite using multi-scale continuum damage model. Materialia, 2019; 6: 100267.
2. Tang, X., et al., Progressive failure analysis of 2×2 braided composite exhibiting multiscale heterogeneity. Composites Science and Technology. 2006; 66(14): 2580–2590.
3. Kanouté, P., et al., Multiscale Methods for Composites: A Archives of Computational Methods in Engineering, 2009; 16(1): 31–75.
4. Cheng, J., et al., A review on the multi-scale simulation of Z-pinned composites laminates. Composite Structures, 2022; 295: 115834.
5. Teng, J., Z. Zhuang, and L. Bintai, A study on low-velocity impact damage of Z-pinned CFRP composites. Journal of Mechanical Science and Technology, 2007; 21: 2125–2132.
6. Meo, M., F. Achard, and M. Grassi, Finite element modelling of bridging micro-mechanics in through-thickness reinforced composite laminates. Composite Structures, 2005; 71(3): 383–387.
7. Nguyen, A.T.T., et al., Failure modelling and characterisation for pin-reinforced metal-composite joints. Composite Structures. 2018. 188: 185–196.
8. Li, M. and P. Chen, A new FE model for predicting the bridging micromechanisms of a Z-pin. Composite Structures. 2019. 223: 110957.
9. Li, M., et al., Experimental and numerical study on the tensile properties of T-joints with low Z-pin volume density. Polymer Composites. 2020. 41(1): 258–270.
10. Cui, H., et al., Mixed mode cohesive law for Z-pinned composites analyses. Computational Materials Science, 2013. 75: 60–68.
11. Melro, A.R., et al., An energy-equivalent bridging map formulation for modelling delamination in through-thickness reinforced composite laminates. International Journal of Solids And Structures. 2020. 202: 153–165.
12. Nguyen, A.T.T., et al., Loading, support and geometry effects for pin-reinforced hybrid metal-composite joints. Composites Part A: Applied Science and Manufacturing. 2017. 98: 192–206.
13. Grassi, M. and X. Zhang, Finite element analyses of mode I interlaminar delamination in z-fibre reinforced composite laminates. Composites Science and Technology. 2003; 63(12): 1815–1832.
14. Camanho, P. and C. Dávila, Mixed-Mode Decohesion Finite Elements for the Simulation of Delamination in Composite Materials. Materials Science, 2002.

15. Heidari-Rarani, M. and A.R. Ghasemi, Appropriate shape of cohesive zone model for delamination propagation in ENF specimens with R-curve effects. Theoretical and Applied Fracture Mechanics. 2017; 90: 174–181.
16. Bianchi, F. and X. Zhang, A cohesive zone model for predicting delamination suppression in Z-pinned CFRP composites. Composites Science and Technology. 2011; 71(16): 1898–1907.
17. Bianchi, F., et al., Finite element modelling of Z-pinned composites T-joints. Composites Science and Technology. 2012; 73: 48–56.
18. Cartié, D.D.R., M. Troulis, and I.K. Partridge, Delamination of Z-pinned carbon fibre reinforced laminates. Composites Science and Technology. 2006; 66(6): 855–861.
19. Mouritz, A.P., Environmental durability of Z-pinned carbon fibre–epoxy laminate exposed to water. Composites Science and Technology, 2012; 72: 1568–1574.
20. Zhang, B., et al., Micro-mechanical finite element analysis of Z-pins under mixed-mode loading. Composites Part A: Applied Science and Manufacturing. 2015; 78: 424–435.
21. Bianchi, F. and X. Zhang, Predicting mode-II delamination suppression in Z-pinned CFRP composites. Composites Science and Technology. 2012; 72(8): 924–932.
22. Barrett, D.J., The mechanics of Z-fiber reinforcement. Composite Structures. 1996; 36(1): 23–32.
23. Kravchenko, S., et al., Composite toughness enhancement with interlaminar reinforcement. Composites Part A: Applied Science and Manufacturing. 2013; 54: 98–106.
24. Zhang, B., G. Allegri, and S. Hallett, Micro-Mechanical Finite Element Analysis of Single Z-pin Pull-out, in SIMULIA UK Regional User Meeting. 2013.
25. Yasaee, M., G. Mohamed, and S.R. Hallett, Interaction of Z-pins with Multiple Mode II Delaminations in Composite Laminates. Experimental Mechanics. 2016; 56(8): 1363–1372.
26. Zi, G. and T. Belytschko, New crack-tip elements for XFEM and applications to cohesive cracks. 2003; 57(15): 2221–2240.
27. Cui, H., et al., Bridging micromechanisms of Z-pin in mixed mode delamination. Composite Structures. 2011; 93(11): 2685–2695.
28. Cartié, D.D.R., et al., 3D reinforcement of stiffener-to-skin T-joints by Z-pinning and tufting. Engineering Fracture Mechanics. 2006; 73(16): 2532–2540.
29. Mouritz, A.P., Review of Z-pinned CFRP composites and sandwich composite. Composites Part A: Applied Science and Manufacturing. 2020; 139: 106128.
30. Nguyen, A.T.T., et al., Pin pull-out behaviour for hybrid metal-composite joints with integrated reinforcements. Composite Structures 2016; 155: 160–172.
31. Yasaee, M., et al., Experimental characterisation of mixed mode traction–displacement relationships for a single carbon composite Z-pin. 2014; 94: 123–131.
32. Allegri, G. and X. Zhang, On the delamination and debond suppression in structural joints by Z-fibre pinning. Composites Part A: Applied Science and Manufacturing. 2007; 38: 1107–1115.
33. Pingkarawat, K. and A.P. Mouritz, Improving the mode I delamination fatigue resistance of composite using Z-pins. Composites Science and Technology, 2014. 92: 70–76.

# Chapter 5
# Modeling of the Influence of Curing Process on Mechanical Properties of Z-pinned CFRP Composites

Residual stress is widely regarded as a crucial factor affecting the manufacturing quality of Z-pinned CFRP composites in the curing process. Varieties of complex physical phenomena existing in the composite curing are attributed to the mismatched thermal expansion and chemical shrinkage between different components at a micro-scale. It is clear that Z-pinning insertion produces several types of fine-scale damage around Z-pins. The developed high-stress concentration exceeds the strength of interface and is responsible for the local cracking in the vicinity of Z-pins after the cure. It can be pointed out that interfacial cracking around Z-pins is an inherent problem during the cure reaction and is essentially unavoidable. The existence of interfacial debonding does not diminish the reported advantages of Z-pinning, as the reinforcement carried out may have been performed on laminates involving cracking in reality.

Z-pinning affecting the in-plane mechanical performance of laminates is attracting considerable critical attention, due to microstructural damage within the base laminates. The presence of Z-pins could produce fiber distortion and resin-rich pockets and lead to a reduction of strength and stiffness of the laminates. The affected damage area increases with increasing Z-pin diameter, and the in-plane tensile properties reduces with increasing Z-pin volume content. The considerable curing mismatch between fiber and matrix may adversely change the microstructures and should not be negligible. The subsequent defects caused by the huge curing mismatch are inherent problems before external loading. In addition, bridging models have been long established in finite element methods to evaluate the inter-laminar fracture toughness of Z-pinned CFRP composites by pullout loading [1]. Since Z-pins act as fine nails to ensure the interlocking between two piles, the phenomenon of interfacial cracking surrounding Z-pins should be of great concern. However, current models are proposed on the basis of several simplifying assumptions. In some studies [2, 3], it is assumed that Z-pins are perfectly bonded to the base laminates after cure. On the contrary, the responses of Z-pins are dominated by fully debonding before external loading applied [4]. Typically, cohesive modeling is adopted to capture the complexities of



Y. Xu and W. Zhang, *Modeling of Z-pinned Carbon Fiber-Reinforced Polymer (CFRP) Composite*, https://doi.org/10.1007/978-981-96-3628-0_5

Z-pin bridging behavior in terms of separation-traction law. These existing bridging models use several simplifying assumptions about the physical interfacial adhesion and failure mode of Z-pins subjected to loading. The random variation of interfacial debonding caused by huge curing residual stresses may not have received systematic attention, which may compromise the accurate prediction of Z-pin reinforcement.

It is important to note that the presence of curing defects, including considerable residual stresses and interfacial debonding, is a fundamental problem for Z-pinned CFRP composites throughout the cure. Published findings on the impact of curing defects on the mechanical properties of Z-pins are scarce. A detailed understanding of the mechanical properties of Z-pins, accounting for curing effects (CE), should be demonstrated.

The aim of this chapter is to provide a more reliable analytical method for assessing the mechanical behavior of individual Z-pins within carbon/epoxy base laminates. A comprehensive link between the formation of curing defects and mechanical performance is proposed. The properties of fiber distortion and resin-rich regions are characterized. The evolutions of cure reaction properties of the Z-pinned CFRP composites are described using aforementioned cure-hardening instantaneously linear elastic (CHILE) model. Then, the predicted residual stress is regarded as a predefined boundary and a multi-step mechanical analysis of the cured Z-pinned CFRP composites is proposed. The influences of volume content and diameter of Z-pins on the elastic modulus, tensile strength, and failure mode are determined. The alteration to these dominant features associated with the CE is discussed. In addition, pullout loading is applied to the model to capture the bridging mechanism of Z-pins after the cure. The simulated load–displacement curves with and without CE are compared. The relationship between the interfacial debonding and maximum pullout force is determined. Furthermore, the impact of curing parameters on the interfacial cracking and bridging tractions is investigated by means of the proposed approach.

## 5.1 Constitutive Modeling of Z-pinned CFRP Composites

### *5.1.1 Resin Matrix*

Resin is often observed to possess isotropic elasticity, yielding, and fracture behaviors when subjected to external loadings. To capture this characteristic, Drucker–Prager criterion [5] is deployed here to describe the yielding, which is governed by:

$$F = t - p \cdot \tan\psi - \kappa = 0 \tag{5.1}$$

where $p = -\frac{1}{3}\mathrm{trace}(\sigma)$ is the equivalent pressure, $t$ is the yield stress, $\psi$ is the friction angle, and $\kappa$ is the resin cohesion which is associated with the uniaxial tension yield stress. The values of $t$ and $\kappa$ can be respectively computed by:

$$t = \frac{1}{2}q\left[1 + \frac{1}{k} - \left(1 - \frac{1}{k}\right)\left(\frac{r}{q}\right)^3\right] \tag{5.2}$$

$$\kappa = \left(\frac{1}{k} + \frac{1}{3}\tan\psi\right)\sigma_t \tag{5.3}$$

where $q$ is the equivalent Mises stress, $k$ is the yield stress ratio, and $r$ is the deviatoric stress tensor.

Ductile criterion is implemented to estimate the damage evolution of the resin. The initiation and progression of the resin damage are certain of the amplitude of an equivalent plastic strain. The distinct phases of the stress–strain response under external loadings coupled with the damage evolution are drawn in Fig. 5.1. This relationship indicates that the damage begins at an equivalent plastic strain value of $\overline{\varepsilon}_0^{pl}$, and the failure is completed when the plastic strain reaches $\overline{\varepsilon}_f^{pl}$.. After point ($\overline{\varepsilon}_0^{pl}$, $\sigma_{y0}$), the dashed curve in Fig. 5.1 shows ideal behavior in absence of the damage, and the solid part depicts effective behavior with the damage into consideration. After damage initiation, the growth of the resin damage is controlled by progressive degradation of material stiffness. A degradation factor $D(\varepsilon)$ is computed to capture the progression, and the definition can be given by:

$$D(\varepsilon) = \frac{\overline{\varepsilon}^{pl}}{\overline{\varepsilon}_f^{pl}} \tag{5.4}$$

In the end, the fracture energy $G_D$ is applied to examine the ultimate failure of the resin when $D(\varepsilon)$ reaches 1.

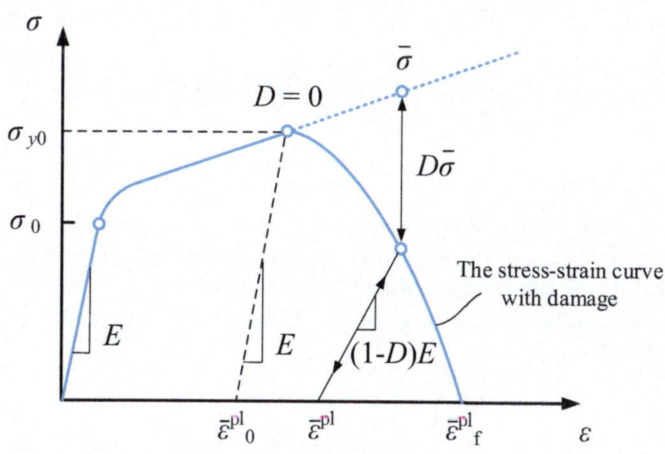


**Fig. 5.1** Stress–strain relationship coupled with damage evolution

### 5.1.2 Z-pin

Maximum nominal stress criterion is utilized to estimate onset of Z-pin damage under loadings. It is assumed that once the damage occurs, the maximum nominal stress ratio reaches 1. Thereby, the expression can be given by:

$$\max\left\{\frac{\langle\sigma_{\text{p,n}}\rangle}{\sigma_{\text{p,n}}^{0}}, \frac{\sigma_{\text{p,s}}}{\sigma_{\text{p,s}}^{0}}, \frac{\sigma_{\text{p,t}}}{\sigma_{\text{p,t}}^{0}}\right\} = 1 \tag{5.5}$$

where $\sigma_{\text{p,n}}{}^{0}$, $\sigma_{\text{p,s}}{}^{0}$ and $\sigma_{\text{p,t}}{}^{0}$ are on behalf of Z-pin tensile strengths in the normal and two orthogonal shear directions, respectively, $\sigma_{\text{p,n}}$, $\sigma_{\text{p,s}}$ and $\sigma_{\text{p,t}}$ represent the tensile stresses in the above three directions, respectively, $\langle\rangle$ illustrates the Macaulay bracket that purely considers tensile state.

### 5.1.3 Interface Between Z-pin and Resin Matrix

Interface is regarded as an attachment between Z-pin and resin-rich pocket parts to transfer complex stresses. CZM is implemented to capture the performance guided by bilinear traction–separation law. The bilinear traction–separation law under mixed-mode condition is depicted in Fig. 5.2. In the beginning, the cohesive response follows an elastic ascending phase without damage. As the loading increases to the interfacial strength, the stiffness begins to degrade, and the damage initiates. Hereafter, the damage evolution is governed by a linear softening law. When the traction ultimately reduces to zero, the load-bearing capacity of the connection disappears and interfacial cracking occurs. The linear elasticity in absence of damage can be provided by:

$$\begin{Bmatrix} \sigma_{\text{n}} \\ \sigma_{\text{s}} \\ \sigma_{\text{t}} \end{Bmatrix} = \begin{Bmatrix} K_{\text{nn}} & 0 & 0 \\ 0 & K_{\text{ss}} & 0 \\ 0 & 0 & K_{\text{tt}} \end{Bmatrix} \begin{Bmatrix} \delta_{\text{n}} \\ \delta_{\text{s}} \\ \delta_{\text{t}} \end{Bmatrix} \tag{5.6}$$

where $\sigma_{\text{n}}$, $\sigma_{\text{s}}$ and $\sigma_{\text{t}}$ are the interfacial traction stresses in the normal and two orthogonal shear directions, respectively, $\delta_{\text{n}}$, $\delta_{\text{s}}$ and $\delta_{\text{t}}$ are the corresponding displacements, $K_{\text{nn}}$, $K_{\text{ss}}$ and $K_{\text{tt}}$ are the corresponding elastic stiffness.

The QUADS stress criterion is utilized to calculate the beginning of the stiffness degradation:

$$\left\{\frac{\langle\sigma_{\text{n}}\rangle}{\sigma_{\text{n}}^{0}}\right\}^{2} + \left\{\frac{\sigma_{\text{s}}}{\sigma_{\text{s}}^{0}}\right\}^{2} + \left\{\frac{\sigma_{\text{t}}}{\sigma_{\text{t}}^{0}}\right\}^{2} = 1 \tag{5.7}$$

where $\sigma_{\text{n}}{}^{0}$, $\sigma_{\text{s}}{}^{0}$ and $\sigma_{\text{t}}{}^{0}$ are the interfacial strengths in the mentioned three directions, respectively.

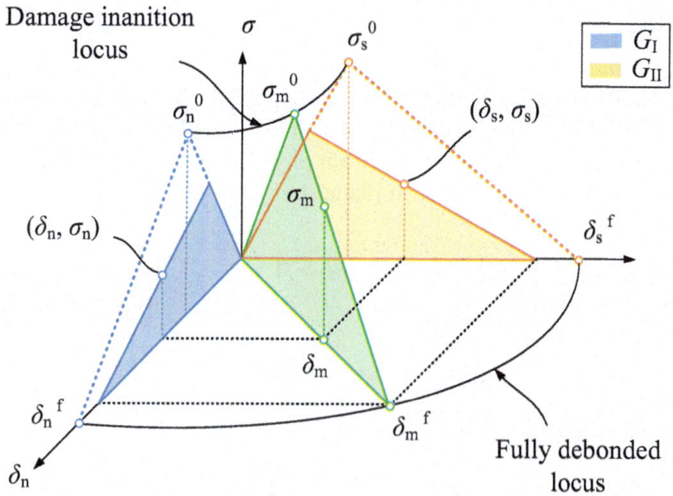


**Fig. 5.2** Mixed-mode constitutive law for interface

After the damage initiation, the traction is governed by the degradation factor $D(\delta_m)$:

$$D(\delta_m) = \frac{\delta_m^f\left(\delta_m^{max} - \delta_m^0\right)}{\delta_m^{max}\left(\delta_m^f - \delta_m^0\right)} \tag{5.8}$$

$$\delta_m = \sqrt{\langle\delta_n\rangle^2 + \delta_s^2} \tag{5.9}$$

where $\delta^0{}_m$ and $\delta^f{}_m$ are the effective separations at the damage initiation and damage completion, respectively, $\delta_m^{max}$ is the effective separation in a maximum value, and $\delta_m$ is the mixed effective separation with a combination of normal section and shear section.

The B–K fracture criterion is introduced to compute the interfacial cracking based on energy dissipated, and the relation can be expressed by:

$$G_c = G_{Ic} + (G_{IIc} - G_{Ic})\left(\frac{G_{II}}{G_I + G_{II}}\right)^{\eta_{BK}} \tag{5.10}$$

where $G_c$ is the total energy release rate, $G_{Ic}$ and $G_{IIc}$ are the two critical fracture toughness, and $\xi$ represents a mixed-mode coefficient.

### 5.1.4 Base CFRP

Hashin's theory is typically applied to determine the failure mechanism for the lamina. Fiber and matrix of the lamina are comprehensively modified in the case of tension and compression based on physical considerations. The analytical descriptions of Hashin criteria for in-plane stress are summarized:

$$\begin{cases} F_{\mathrm{ft}} = \left(\frac{\sigma_{11}}{X_{\mathrm{T}}}\right)^2 + \left(\frac{\sigma_{12}}{S_{\mathrm{L}}}\right)^2 = 1, \sigma_{11} > 0 \\ F_{\mathrm{fc}} = \left(\frac{\sigma_{11}}{X_{\mathrm{C}}}\right)^2 = 1, \sigma_{11} < 0 \\ F_{\mathrm{rt}} = \left(\frac{\sigma_{22}}{Y_{\mathrm{T}}}\right)^2 + \left(\frac{\sigma_{12}}{S_{\mathrm{L}}}\right)^2 = 1, \sigma_{22} > 0 \\ F_{\mathrm{rc}} = \left(\frac{\sigma_{22}}{2S_{\mathrm{T}}}\right)^2 + \left[\left(\frac{Y_{\mathrm{C}}}{2S_{\mathrm{T}}}\right)^2 - 1\right]\frac{\sigma_{22}}{Y_{\mathrm{C}}} + \left(\frac{\sigma_{12}}{S_{\mathrm{L}}}\right)^2 = 1, \sigma_{22} < 0 \end{cases} \tag{5.11}$$

where $F$ denotes the failure index, the subscripts t and c indicate tension and compression, $\sigma_{11}$, $\sigma_{22}$ and $\sigma_{12}$ represent the normal or shear stresses, $X_{\mathrm{T}}$, $X_{\mathrm{C}}$, $Y_{\mathrm{T}}$ and $Y_{\mathrm{C}}$ are the longitudinal and transverse strengths under tension and compression, $S_{\mathrm{L}}$ and $S_{\mathrm{T}}$ are the longitudinal and transverse shear strengths, respectively.

## 5.2 In-Plane Mechanical Properties

### 5.2.1 Geometrical Shaping Operations of Resin-Rich Pocket

Boundary of the resin-rich pocket is mostly assumed as a cosines shape or a diamond shape in the previous analytical research [6]. However, another in-plane border of the resin-rich pocket is presented in Hoffmann's work [7]. Accounting for this, a concave shape is yielded in this model to represent its reasonable presence. To summarize, three geometrical shapes of the resin-rich pocket used in this work are shown in Fig. 5.3. An opening distance of 4 μm is added in the resin-rich location to avoid computational distortion of the pocket at sharp corners. An opening distance of 4 μm is added in the resin-rich location to avoid computational distortion of the pocket at sharp corners. The geometry of the RVE model for inserting thin and thick Z-pins is respectively listed in Table 5.1.

The effect of the Z-pin volume content on the form of resin-rich pocket is also considered during the RVE generation. The Z-pin content can be calculated via the following expression:

$$V_{\mathrm{pin}} = \frac{\pi d^2}{4S^2} \tag{5.12}$$

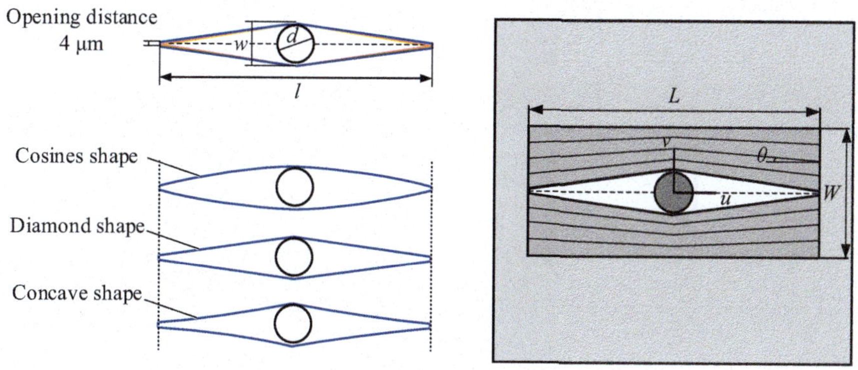


**Fig. 5.3** Three shapes of the resin-rich pocket in the RVE model

**Table 5.1** Geometrical parameters of the RVE for inserting thin and thick Z-pins

| | $d$ (mm) | $L$ (mm) | $W$ (mm) | $l$ (mm) | $w$ (mm) | $\theta_{max}$ (°) |
|---|---|---|---|---|---|---|
| Thin Z-pin | 0.28 | 2.0 | 0.8 | 2.0 | 0.30 | 4.0 |
| Thick Z-pin | 0.51 | 2.8 | 1.5 | 2.8 | 0.53 | 5.4 |

where $d$ is the Z-pin diameter, S is the spacing between two neighboring Z-pins. Significantly, when the spacing is relatively short, a resin-rich channel will be developed. The microstructural morphologies of the resin-rich pockets formed at low level and high level of the Z-pin contents are represented in Fig. 5.4 [8].

### 5.2.2 *Modeling of the Cure Reaction*

Cure reaction in an autoclave is a complexed process for composite structures which contains a series of thermal-chemical consequences under the predefined temperature and pressure profiles. The thermosetting resin is transformed from a viscous liquid into a hard glassy solid in the cure, which is accompanied by significant changes in material properties. A standard differential scanning calorimetry (DSC) test is required to characterize the cure reaction, and a reasonable cure kinetics model is proposed to capture the reaction mechanism for the resin systems.

Degree of cure (DoC), acting as a critical measurement, is introduced to describe the extent of cure reaction by the DSC test results. The expression of DoC based on the heat release of bond formation is given by:

$$\alpha = \frac{1}{H_r}\int_0^t \left(\frac{\mathrm{d}q}{\mathrm{d}t}\right)\mathrm{d}t \tag{5.13}$$

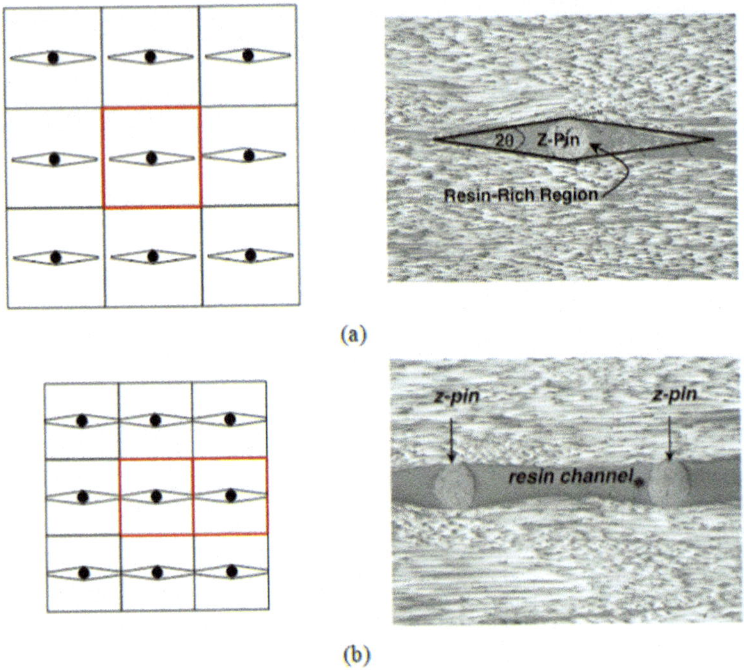


**Fig. 5.4** Resin-rich regions around Z-pins: **a** low level of Z-pin volume content, **b** high level of Z-pin volume content

where $\alpha$ is the DoC, $\frac{dq}{dt}$ stands for the instantaneous rate of heat generation, $H_r$ is the total amount of heat release, and $t$ is the curing time.

An autocatalytic model is applied in this study to reflect the nonlinear relationship between curing rate and its control parameters during cure [9]. The cure kinetic equation for epoxy matrix can be defined as follows:

$$\frac{d\alpha}{dt} = K\alpha^m(1-\alpha)^n \tag{5.14}$$

where $\frac{d\alpha}{dt}$ is the instantaneous curing rate, $m$ and $n$ are the experimentally-determined reaction orders, and $K$ is the constant of curing rate determined by the Arrhenius-type equation:

$$K = A\exp(-\Delta E/RT) \tag{5.15}$$

where $A$ is the pre-exponential constant, $\Delta E$ is the cure activation energy, $R$ is the universal gas constant, and $T$ is the absolute curing temperature.

An instantaneous linear constitutive model [10] is taken here to characterize the autoclave processing phenomena, like developments of curing residual stresses and curing deformations. Thermal expansion and chemical shrinkage acting as contributing factors are considered to describe such evolutions for consistency [11]. Hence, the definition of residual stress $\boldsymbol{\sigma}$ can be expressed as follows:

$$\{\sigma\}=[\mathbf{C}](\{\varepsilon\}-\{\varepsilon_t\}-\{\varepsilon_c\}) \tag{5.16}$$

where $\varepsilon$, $\varepsilon_t$, and $\varepsilon_c$ stand for the total, thermal and chemical strains, respectively, $\mathbf{C}$ is the stiffness tensor which is governed by the elastic modulus of the chosen material.

The isotropic epoxy resin is assumed as a cure-hardening instantaneously linear elastic (CHILE) material following the idea of [10], and the equation of $E_r$ can be formulated as:

$$E_r=\begin{cases} E_r^0 & T^*<T_{C1}^* \\ E_r^0+\frac{T^*-T_{C1}^*}{T_{C2}^*-T_{C1}^*}\left(E_r^\infty-E_r^0\right) & T_{C1}^*\le T^*<T_{C2}^* \\ E_r^\infty & T_{C2}^*\le T^* \end{cases} \tag{5.17}$$

where $E_r{}^0$ and $E_r{}^\infty$ represent the fully uncured and fully cured elastic moduli, respectively. The relationship of these two moduli is $E_r{}^0=E_r{}^\infty/1000$. $T^*$ denotes the difference between instantaneous curing temperature and resin glass transition temperature. $T_{C1}{}^*$ and $T_{C2}{}^*$ are the critical constants at the initiation and completion of the resin glass transition.

The lamina behaves as a transversely isotropic material and its mechanical properties can be defined by the constants of fiber and resin components in terms of the micromechanics model [12]. The in-plane mechanical properties (elastic modulus $E$, shear modulus $G$, and Poisson's ratio $\nu$) are defined using a rule of mixture (RoM) approach [13] and given by:

$$\begin{cases} E_1=E_{f1}V_f+E_r(1-V_f) \\ E_2=\dfrac{E_{f2}E_r}{E_{f2}(1-V_f)+E_rV_f} \\ G_{12}=\dfrac{G_{f12}G_r}{G_{f12}(1-V_f)+G_rV_f} \\ \nu_{12}=\nu_{f12}V_f+(1-V_f)\nu_r \end{cases} \tag{5.18}$$

where the subscripts 1 and 2 are the longitudinal and transversal directions of the lamina, f and r stand for the fiber and resin, respectively.

The thermal strain $\varepsilon_t$ is a function of coefficient of thermal expansion (CTE) and curing temperature, which is formulated as:

$$\varepsilon_t=\beta\cdot T \tag{5.19}$$

where $\beta$ stands for the CTE which is related to the DoC evolution. The CTE for the resin can be calculated using the RoM approach as follows:

$$\beta_r = (1-\alpha)\beta_r^r + \alpha\beta_r^g \tag{5.20}$$

where $\beta_r^r$ and $\beta_r^g$ are respectively determined by the constants of rubbery state and glassy state during the cure process.

For the aspect of the lamina, the integrated CTEs in the longitudinal and transverse directions can be determined by a self-consistent scheme and thereby show fairly accuracy by the following relations:

$$\begin{cases} \beta_1 = \frac{\beta_{f1}E_{f1}V_f + \beta_r E_r(1-V_f)}{E_{f1}V_f + E_r(1-V_f)} \\ \beta_2 = (\beta_{f2} + v_{f12}\beta_{f1})V_f + (\beta_r + v_r\beta_r)(1-V_f) \\ \quad -[v_{f12}V_f + v_r(1-V_f)]\beta_1 \end{cases} \tag{5.21}$$

The chemical strain $\varepsilon_c$ is identified by the coefficient of chemical shrinkage (CCS) and progression of DoC. According to Abou msallem et al. [14], the expression is written as:

$$\varepsilon_c = \gamma \cdot \alpha \tag{5.22}$$

where $\gamma$ denotes the CCS.

The expressions of the homogenized CCSs for the lamina are derived using the identical form as Eq. (5.23). A zero CCS is set here for the fiber component due to its minor impact on the overall chemical shrinkage. Consequently, the integrated CCSs in the two principal directions of the lamina can be simplified as below:

$$\begin{cases} \gamma_1 = \frac{\gamma_r E_r(1-V_f)}{E_{f1}V_f + E_r(1-V_f)} \\ \gamma_2 = (1+v_r)(1-V_f)\gamma_r - [v_{f12}V_f + v_r(1-V_f)]\gamma_1 \end{cases} \tag{5.23}$$

Z-pin is assumed to be a fiber-dominant material, and the thermal expansion and chemical shrinkage compared to other components are negligible during cure.

### 5.2.3 *Finite Element Implementation*

A two-dimensional (2D) representative volume element (RVE) of Z-pinned CFRP composites defined in this work is taken directly from the experimental studies conducted in previous publications. The curing and subsequent in-plane tensile loading in the Z-pinned CFRP composites are systematically simulated utilizing the RVE-based numerical approach. As depicted in Fig. 5.5a, a two-step coupled computation is developed: Step I analyses the cure reaction processing, and Step II estimates the tensile performance after the cure. Especially, the thermal-chemical evolutions during the curing are characterized by a combination of UEXPAN and UMAT. The obtained cure-induced residual stresses from Step 1 acting as a significant role are applied as a predefined boundary in the subsequent in-plane tension analysis. To demonstrate the accuracy of the modeling strategy proposed, the in-plane tensile parameters of the cured Z-pinned specimens are also taken from Chang et al. [15]. The curing temperature and pressure profiles performed in Chang's experiments are drawn in Fig. 5.5b. The material of the lamina is selected using T300/CYCOM 970 prepregs and the Z-pin is T300/BMI. Refer to the database provided by CYTEC Company and literature, the curing parameters of the chosen materials are listed in Tables 5.2 and 5.3 in great detail [16]. Moreover, the typical numerical parameters input in the constitutive modeling of the resin, interface, and lamina are given in Table 5.4, in order to describe the in-plane tensile behavior. As shown in Fig. 5.5c, the 2D mesoscopic modeling of the Z-pinned CFRP composites is constructed based on the description in Chap. 3. A layer of quadrilateral elements (CPS4R) is utilized for the lamina. Cohesive elements (COH2D4) are implemented to realize the interfacial behavior. Via mesh consideration, the mesh size of the RVE is determined to be 0.08. Periodical boundary conditions are introduced to maintain the continuity between neighboring RVEs.

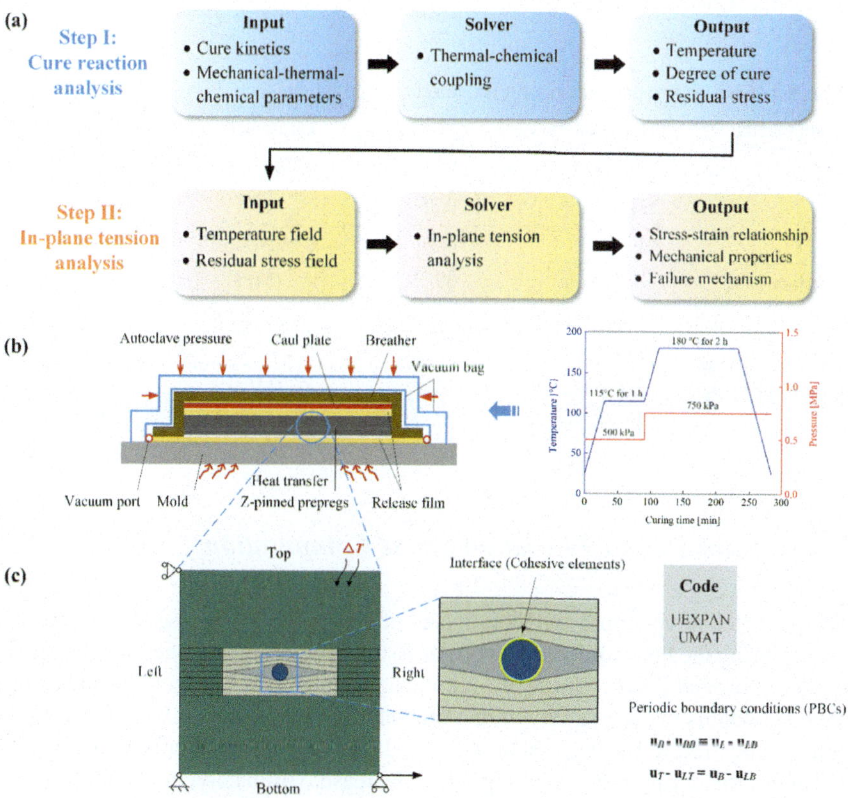


**Fig. 5.5** Flow chart of the in-plane tension analysis considering curing effects: **a** a linking of curing and in-plane tensile analysis, **b** applied curing temperature and pressure profiles, **c** 2D mesoscopic modeling

**Table 5.2** Cure kinetic coefficients of CYCOM 970 epoxy

| Parameters | $A$ ($s^{-1}$) | $E$ (J/mol) | $H_r$ (J/kg) | $m$ | $n$ | $R$ (J $mol^{-1}$ $°C^{-1}$) |
|---|---|---|---|---|---|---|
| Values | 127,000 | 63,400 | 483,000 | 0.9 | 2.1 | 8.314 |

**Table 5.3** Curing properties of T300/CYCOM 970 lamina reinforced with T300/BMI Z-pin

| Properties | Fiber | Resin | Z-pin |
|---|---|---|---|
| Longitudinal modulus (MPa) | $E_{f1} = 230{,}000$ | $E_r^0 = 3.35$ | $E_1 = 100{,}000$ |
| Transverse modulus (MPa) | $E_{f2} = 13{,}800$ | $E_r^\infty = 3350$ | $E_2 = 30{,}000$ |
| Shear modulus (MPa) | $G_{f12} = 8970$ | $G_r = 1240$ | $G_{12} = 4600$ |
| Poisson's ratio | $\nu_{f12} = 0.2$ | $\nu_r = 0.35$ | $\nu_{12} = 0.35$ |
| Coefficient of thermal expansion ($°C^{-1}$) | $\beta_{f1} = -0.7 \times 10^{-6}$ | $\beta_r^u = 81.4 \times 10^{-6}$ | 0 |
| Coefficient of thermal expansion ($°C^{-1}$) | $\beta_{f2} = 12 \times 10^{-6}$ | $\beta_r^c = 34.6 \times 10^{-6}$ | 0 |
| Coefficient of chemical shrinkage | 0 | $\gamma_r = -0.02$ | 0 |

**Table 5.4** Constitutive parameters of resin, interface, and lamina

| Components | Parameters | Values | Parameters | Values |
|---|---|---|---|---|
| Resin [17] | $\kappa$ (MPa) | 97.6 | $\psi$ (°) | 37.7 |
| | $\bar{\varepsilon}_{0-}^{pl}$ | 0.25 | $\bar{\varepsilon}_{0+}^{pl}$ | 0.025 |
| | $k$ | 0.8 | $G_D$ (N/mm) | 0.001 |
| Interface [18] | $\sigma_n^0$ (MPa) | 30 | $\sigma_s^0$ (MPa) | 60 |
| | $G_{Ic}$ (N/mm) | 0.1 | $G_{IIc}$ (N/mm) | 0.1 |
| | $\eta_{BK}$ (–) | 1 | | |
| Lamina [19] | $X_T$ (MPa) | 1500 | $X_C$ (MPa) | 900 |
| | $Y_T$ (MPa) | 27 | $Y_C$ (MPa) | 200 |
| | $S_L$ (MPa) | 80 | $S_T$ (MPa) | 41 |

### *5.2.4 Cure-Induced Residual Stress Computation*

The predicted temperature and DoC curves of the RVE are expressed in Fig. 5.6. Up to and during the first isothermal dwell (0–90 min), the resin solidifies slowly and the DoC is improved slowly. During the second heating period of the cure (90–112 min), the resin begins to cure rapidly with a dramatic improvement of the DoC when the curing temperature reaches about 157 °C. In the second isothermal dwell (112–231 min), the curing rate shows a gradual decrease. Ultimately, the DoC maintains about 0.96.

The curing properties of each constituent in the RVE are altered in terms of the ambient temperature and DoC evolutions. The effect of resin-rich geometries

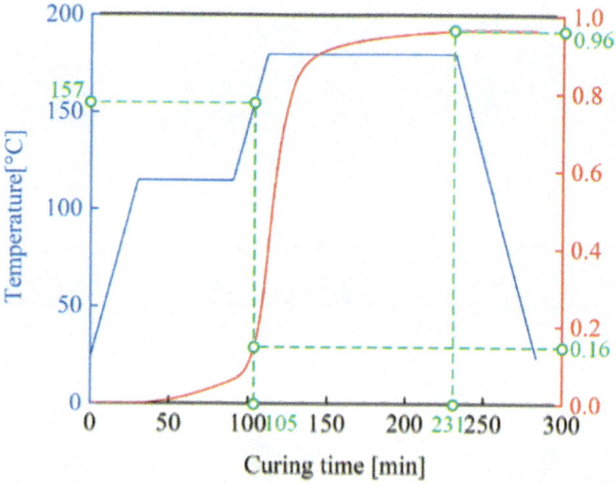


**Fig. 5.6** The predicted temperature and DoC

at low Z-pin content (0.5%) on the curing residual stress distribution is shown in Fig. 5.7. Irrespective of resin pocket geometries, the high stress concentration is developed in the interface, indicating a critical mismatch generated between Z-pin and resin-rich regions during the cure (see Fig. 5.7a–c). The maximum value of Mises residual stress is computed as 79.53 MPa, 71.69 MPa, and 71.20 MPa for the resin pockets represented by the cosine-shape, diamond-shape, and concave-shape, respectively. The usage of cosine-shape accumulates the highest stress concentration while the concave-shape produces the lowest stress concentration. It is believed that the curing residual stresses are suppressed effectively with decreased resin-rich regions. This occurs because the lower resin content facilitates a reduction of swelling as well as curing mismatch surrounding the Z-pins. The maximum principal residual stresses especially for the local fiber distortion around Z-pins are shown in Fig. 5.7d–f. Significantly, the Z-pin experiences a compressive status, the resin-rich pocket exhibits a tensile status, and the fiber waviness nearest the Z-pin shows a distinct compressive status, no matter what the resin pocket geometry is. Moreover, the maximum value of each constituent (interface, Z-pin, resin-rich pocket, and base lamina) is compared in Fig. 5.7g–i.

To understand the transform of Z-pinning on the curing residual stress generation, both maximum values and distributions of the estimated stress for different Z-pin contents and Z-pin diameters are given in Fig. 5.8. The residual stress concentration significantly improves with increasing Z-pin density. The high level of Z-pin density leads to resin-rich channels and thereby more severe fiber squeezing surrounding the Z-pins, owing partly to the resin-rich swelling and partly to the fiber misalignment.

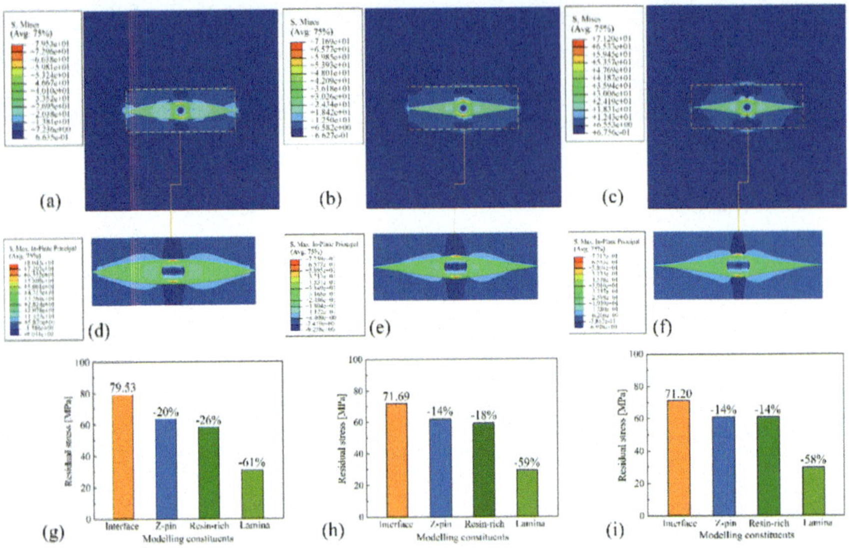


**Fig. 5.7** The effect of resin-rich geometry at low Z-pin content (0.5%) on the curing residual stresses: **a–c** Mises stress, **d–f** maximum principal stress in the local fiber distortion, **g–i** maximum value in different constituents

### 5.2.5 *In-Plane Tensile Properties*

The estimated curing residual stresses are used as the predefined stress fields to further explore the in-plane tensile behavior of the cured Z-pinned CFRP composites. Quasi-static simulation at a strain rate of 1 mm/min based upon Chang experiments has been carried out in ABAQUS explicit. Following volumetric homogenization techniques [20], the equivalent stresses $\overline{\sigma}_{ij}$ and strains $\overline{\varepsilon}_{ij}$ are computed from the RVE modeling, as given by:

$$\overline{\sigma}_{ij} = \frac{1}{V}\int\limits_{V} \sigma_{ij} \mathrm{d}V \tag{5.24}$$

$$\overline{\varepsilon}_{ij} = \frac{1}{V}\int\limits_{V} \varepsilon_{ij} \mathrm{d}V \tag{5.25}$$

where, $\sigma_{ij}$ and $\varepsilon_{ij}$ are the stress and strain components, respectively, and $V$ is the volume of the RVE.

The estimated curing residual stresses are used as the predefined stress fields to further investigate the in-plane tensile behavior of the cured Z-pinned CFRP composites. Quasi-static simulation at a strain rate of 1 mm/min based upon Chang experiments has been carried out in ABAQUS explicit. The results obtained from the

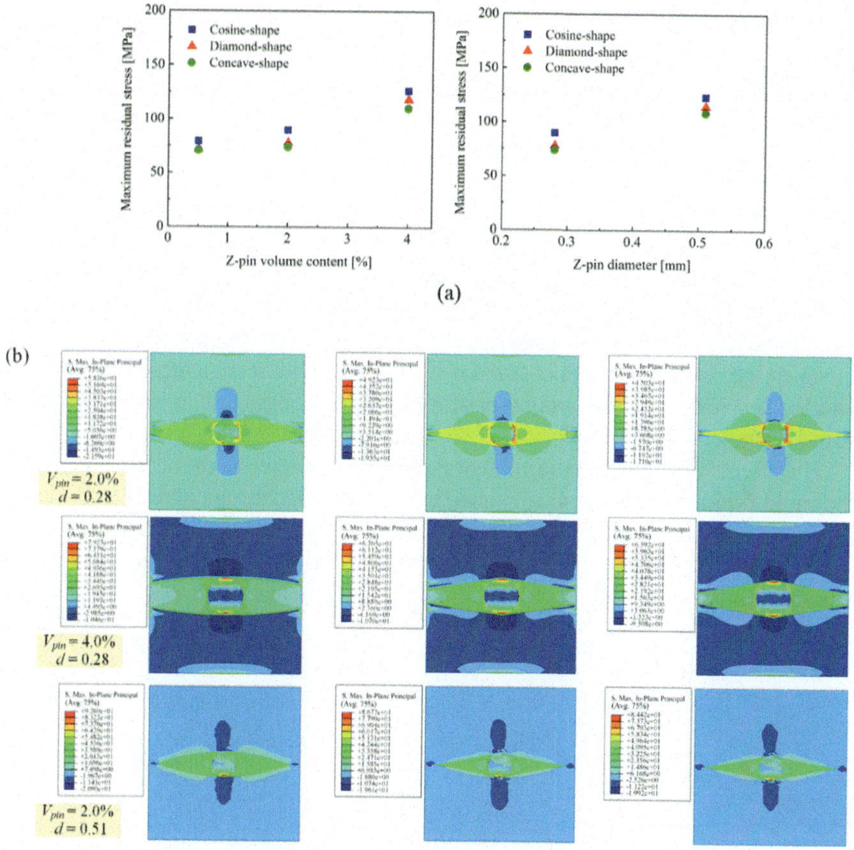


**Fig. 5.8** The effect of Z-pinning parameters on the **a** maximums and **b** distributions of the residual stress of the laminates

simulation of 0.28 Z-pinned CFRP composites are discussed first. The predicted stress–strain relationship with and without curing residual stresses in the case of 0.5, 2.0 and 4.0% Z-pin volume contents is drawn, and the schematic of the force state under tensile loading is shown below corresponding situations, as depicted in Fig. 5.9. The abbreviation CRS in the figure stands for the curing residual stress. Obviously, the effect of CRS on the ultimate failure strain together with tensile strength is more pronounced in the higher Z-pin density (2.0 and 4.0%) than it is for the lower Z-pin density (0.5%). In the case of 0.5% volume content, both strength and failure strain (1576.69 MPa, 0.012) are slightly higher than the data (1507.12 MPa, 0.011) which are not introduced the CRS. In contrast, the estimated strength and failure strain are seriously suppressed for the 2.0 and 4.0% volume contents, with the CRS into consideration. It is worth noting that the curing residual stresses bring about a slight beneficial effect for the in-plane tensile performance of the laminates at low Z-pin density, while lead to a negative impact at intermediate and high Z-pin densities.

This difference may be explained by the enhanced amount of the resin-rich regions formed surrounding the Z-pins. The low Z-pin density produces less resin-rich region and limited swelling during the cure process as mentioned, and thus, the in-plane mechanical behavior of the Z-pinned CFRP composites could be dominated by the unpinned regions. Owing to great thermal expansion along the transverse direction of the lamina, the unpinned region exhibits a compressive state in the longitudinal direction and a tensile state in the transverse direction after the cure. Subsequently, the applied in-plane tensile loading first offset the initial cure-induced states and thus put off the ultimate failure. On the contrary, the high Z-pin density develops continuous resin channels, and a local tensile state forms in the longitudinal direction of the lamina after the cure. In addition, the extremely poor mechanical properties of the interface caused by residual stress concentration with increasing Z-pin density is another key reason for the reduction of the tensile strength and failure strain of the cured Z-pinned CFRP composites.

The effects of Z-pin volume content and Z-pin diameter on the tensile strength of the cured Z-pined laminates are depicted in Fig. 5.10. Experimental results conducted by Chang are compared here to confirm the accuracy of the simulated results obtained in this work. Significantly, the strength decreases rapidly with both increasing Z-pin volume content and Z-pin diameter, and the trends seem to be linear. The values of

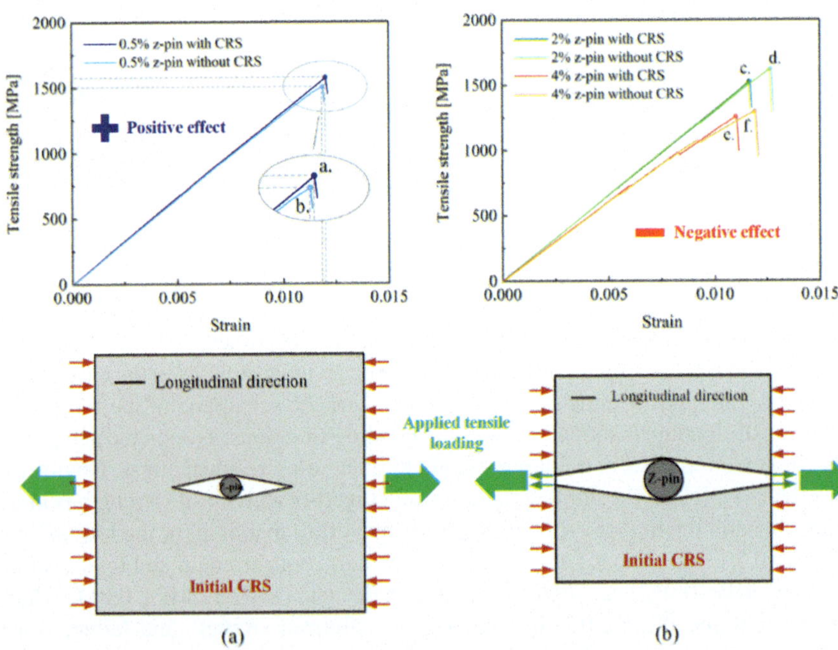


**Fig. 5.9** Stress–strain relationship and force state of the cured Z-pinned CFRP composites at **a** low level of Z-pin density and **b** high level of Z-pin density

strength decrease by approximately 8% and 12% for each 2% Z-pin volume content with embedding thin (0.28 mm) and thick (0.51 mm) Z-pins, respectively.

The effects of Z-pin volume content and Z-pin diameter on the tensile elastic modulus of the cured Z-pined laminates are presented in Fig. 5.11. The effective elastic modulus values are determined using the volumetric homogenization method. The fitted trend lines show linear regressions for either the increasing Z-pin content or the increasing Z-pin diameter. As expected, the laminates reinforced with Z-pins show a reduced elastic modulus, compared to the unpinned laminates. In addition to the fiber distortion, the presence of the CRS weakening the mechanical properties around the Z-pins is responsible for a certain part of the determined degradation in the elastic modulus.

The failure mechanisms at the low, intermediate, and high Z-pin densities of the laminates are compared in Fig. 5.12a–c. As mentioned, the high stress concentration primarily occurs in the interface after the cure. Accordingly, it may be the potential

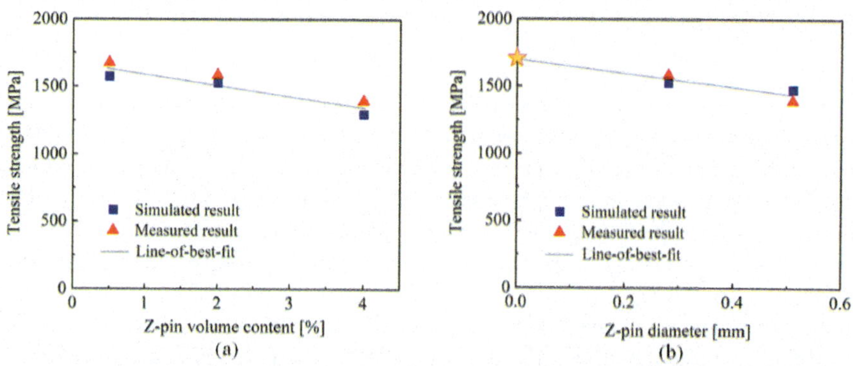


**Fig. 5.10** The effect of **a** Z-pin volume content and **b** Z-pin diameter on the tensile strength of the laminates

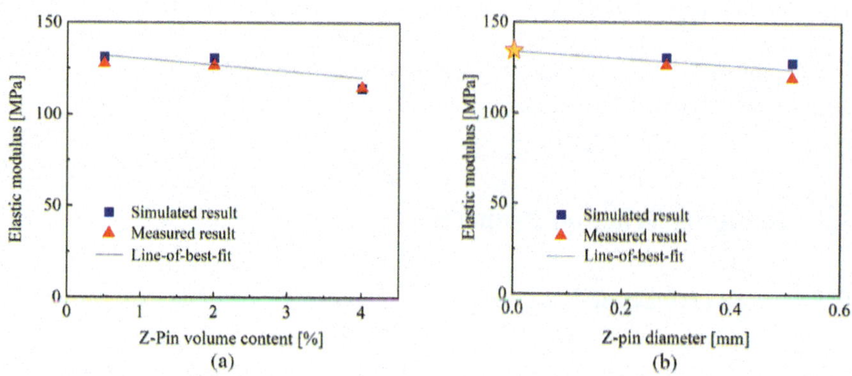


**Fig. 5.11** The effect of **a** Z-pin volume content and **b** Z-pin diameter on the elastic modulus of the laminates

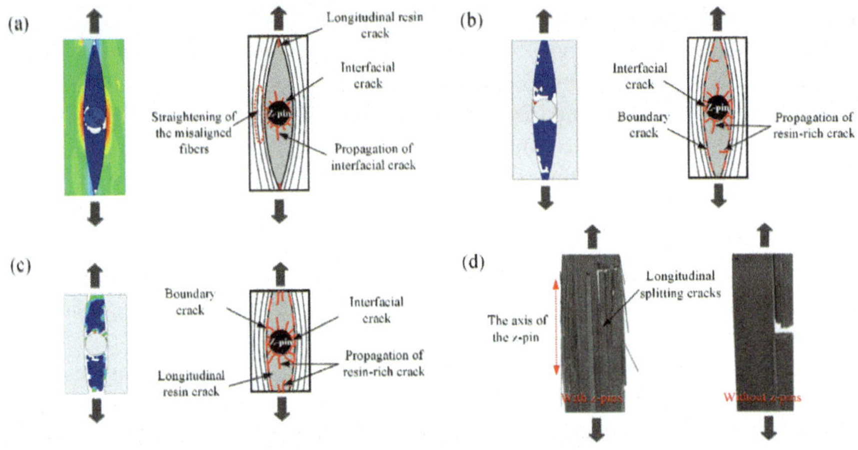


**Fig. 5.12** Tensile failure mechanism in the laminates at **a** low Z-pin density, **b** intermediate Z-pin density, **c** high Z-pin density, and **d** photographs of failed tensile specimens

path for the crack growth under the external loadings. The subsequent tensile loading straightens the misaligned fibers around the Z-pins and promotes the progression of the cracks in the neighboring resin-rich pockets. The 2.0 and 4.0% Z-pin volume contents lead to worsening microstructural defects, including fiber distortion, fiber breakage, resin-rich channels, and swelling, compared to the 0.5% content. Thereby, in addition to the interfacial cracking, the boundary of the resin-rich regions becomes weakened at increased Z-pin density, and resin-rich damages progress seriously when subjected to tensile loading. The change of this failure mode is indeed the major factor in softening the in-plane loading capacity. To verify the predicted mechanism findings, the exanimation of failed test specimens with and without Z-pins subjected to tension is also given in Fig. 5.12d. Obviously, several longitudinal splitting cracks are prompted within the laminates through the axis of the Z-pins. However, in the unpinned laminates, the incidence of the splitting is significantly less. Therefore, it is pointed out that both interfacial stress concentration and resin-rich channels have a strong influence on the growth of cracks, and thereby are primarily responsible for the presence of a load drop driven by tension.

## 5.3 Bridging Traction Properties

### *5.3.1 Experimental Program*

All specimens are manufactured using a unidirectional carbon/epoxy (T300/8552) prepreg reinforced with Polyimide Z-pins supplied by Daober Composite Co., Ltd. Approximately 1000 and 2000 individual fibers are pultruded to fabricate two kinds of Z-pins. The diameter of each kind is measured to be 0.38 mm and 0.55 mm, respectively. The nominal dimension of each specimen is designed as 20 mm × 20 mm. The prepreg plies are arranged in $[0]_{32}$, $[0]_{64}$ and $[0]_{128}$ configurations, respectively. A PTFE release film is inserted at the mid-plane of each stacking sequence, in order to avoid interpenetration and ensure bridging tractions obtained completely from the Z-pins during pullout testing.

A schematic diagram of fabricating process of the Z-pinned specimens is systematically shown in Fig. 5.13. Pre-Hole Z-pinning (PHZ) technique is utilized in this study for controlling a wonderful insertion of the Z-pins. It involves two steps: (1) the lay-up prepregs are heated up and penetrated by a sharp steel needle following the through-thickness direction to create pin-insertion holes; (2) the prepared Z-pins are implemented into the punched holes in a gentle manner and subsequently cut to length (i.e., the nominal thickness of each specimen) by cutter. The detailed description of the PHZ approach can be found in Ref. [21]. Compared to the traditionally Ultrasonically-Assisted Z-pinning (UAZ) method, the attractive feature of the PHZ is minimizing initial microstructural damage within the base laminate, such as fiber crimp, fiber breakage and Z-pin misalignment. Afterward, the Z-pinned prepregs are placed in an autoclave under recommended high temperature and pressure cycles. A growing traction force can be generated between the Z-pin and surrounding base laminate, thereby responsible for locking the piles together after the cure.

Pull-out tests are conducted to assess principal factors that affect the bridging traction behavior of Z-pins after the cure process. A 5-kN-capacity universal testing machine is carried out. The loading rate controlled is 0.5 mm/min. A couple of T-shaped steel pieces are glued to per test specimen and hereafter gripped by the

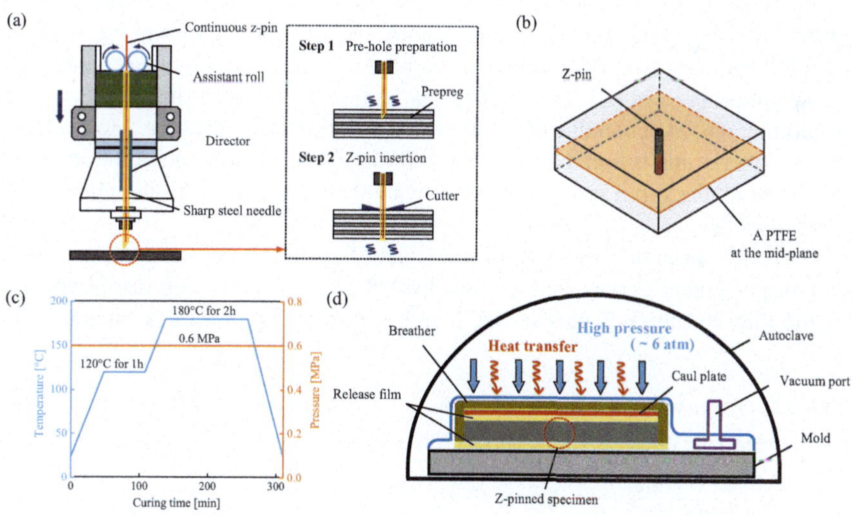


**Fig. 5.13** Schematic illustration of fabricating Z-pinned specimens. **a** Two-step of the PHZ method. **b** Specimen arrangement. **c** Recommended curing cycles. **d** Boundary conditions under autoclaving system

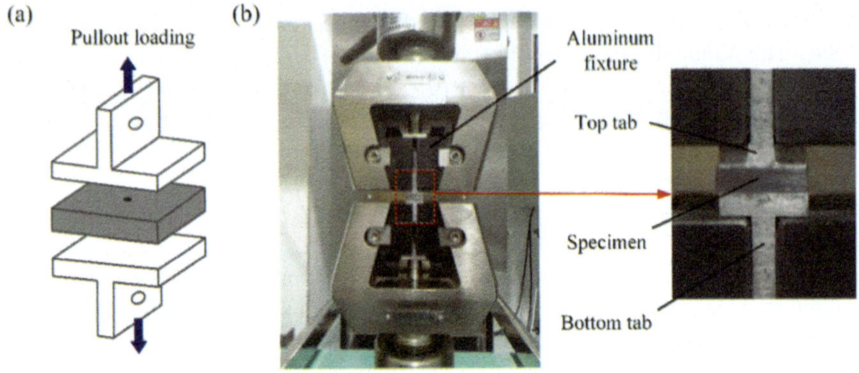


**Fig. 5.14** **a** Loading conditions on Z-pinned specimens; **b** pullout test equipment

machine fixtures during testing, as illustrated in Fig. 5.14. The specific arrangements of the Z-pinned specimens are summarized in Table 5.5. Because of scatter, three samples are tested for each configuration.

SEM approach is conducted before testing to characterize the microstructural damage in the vicinity of Z-pins. The in-situ morphology is presented in Fig. 5.15. Obviously, Z-pins push laminate fibers aside to develop in-plane fiber waviness and become central parts of eyelet-shape resin-rich regions. The size of the eyelet-shaped region directly depends on the diameter of the Z-pin $d$. As depicted in Fig. 5.15a–b, the length of the resin-rich region $l$ is measured to be approximately 5 times of the thin-diameter (0.38 mm) and 8 times of the thick-diameter (0.55 mm), respectively. The in-plane fiber waviness angle $\theta$ is obtained to be 8.6° for the thin Z-pin and 11.3° for the thick Z-pin, respectively. At the higher level of details shown in Fig. 5.15c–e, the interfacial cracking is observed in the boundary connection between the Z-pin and neighboring resin-rich, from both cross-sectional and side-sectional views. This reveals that partial debonding of the Z-pin exists indeed after the cure. Moreover, no obvious fiber crimp is found along the axis of the Z-pin. On account of the advanced PHZ method, the Z-pin shows a quite vertical manner with less misalignment in the base laminate.

The results from the pullout testing are quantified in terms of one crucial relation: load–displacement curves. The pullout load $\boldsymbol{P}$ and corresponding displacement $\delta$ for the thin and thick Z-pins in the 2 mm, 4 mm and 8 mm thick laminates are

**Table 5.5** Overview of the Z-pinned specimens

| No | Stacking sequence | Z-pin embedded length (mm) | Z-pin diameter (mm) |
|---|---|---|---|
| 1 | $[0_{16}/\text{PTFE}/0_{16}]$ | 2 | 0.38 |
| 2 | $[0_{32}/\text{PTFE}/0_{32}]$ | 4 | 0.38 |
| 3 | $[0_{64}/\text{PTFE}/0_{64}]$ | 8 | 0.38 |
| 4 | $[0_{16}/\text{PTFE}/0_{16}]$ | 2 | 0.55 |

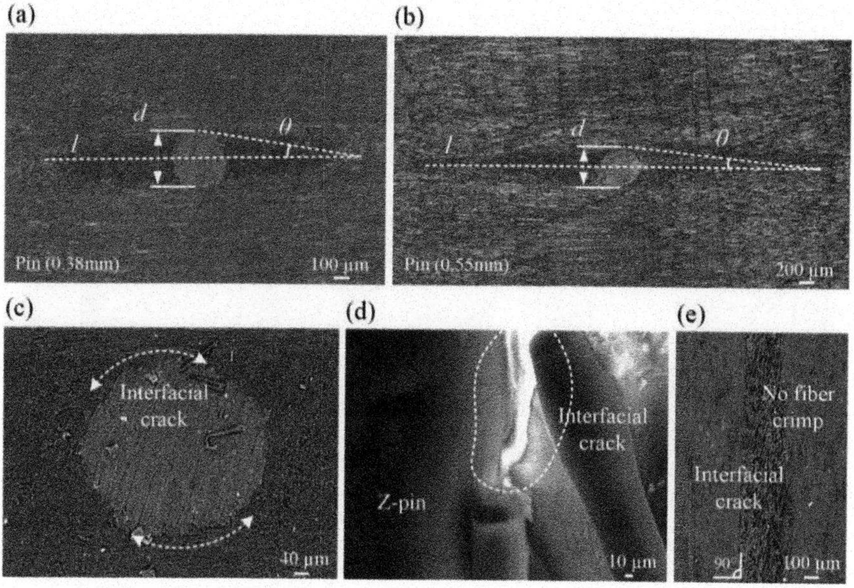


**Fig. 5.15** Microstructural morphologies of **a** thin (0.38 mm) and **b** thick Z-pins (0.55 mm) embedded in carbon/epoxy laminates. **c**–**d** cross-sectional and **e** side-sectional views

plotted in Fig. 5.16a–d, respectively. It is worth noting that all determined load–displacement curves can be characterized by two main stages: (1) at the beginning, the pullout load increases sharply with increasing applied displacement; (2) after reaching a peak value, the pullout load decreases progressively until the Z-pin is entirely pulled out. The final displacements for different conditions are recorded, being approximately half-length of the corresponding thickness of the specimens. In addition, the maximum pullout load $\boldsymbol{P}_{\max}$ and $\boldsymbol{P}_{\max}$ per unit embedded length are collected to identify the effectiveness of Z-pin bridging performance. As shown in Fig. 5.16e–f, the maximum value increases greatly with increasing Z-pin embedded length and Z-pin diameter. This is because the enhanced Z-pin embedment brings about an improvement to bonded connection area, thus benefiting the overall traction capability. In contrast, the $\boldsymbol{P}_{\max}$ per unit embedded length decreases steadily with increasing embedded lengths while increases significantly with increasing Z-pin diameter. This discloses that the increase of Z-pin embedded length remarkably limits the bridging performance per unit embedded length.

To gain insight into the traction mechanism, a bi-linear softening law and corresponding pullout behavior illustration are summarized in Fig. 5.17. Two crucial relation parameters, i.e., the maximum pullout force $\boldsymbol{P}_{\max}$ and its corresponding displacement $\delta_e$ are respectively determined, and consequently, a recommended bridging law can be described by:

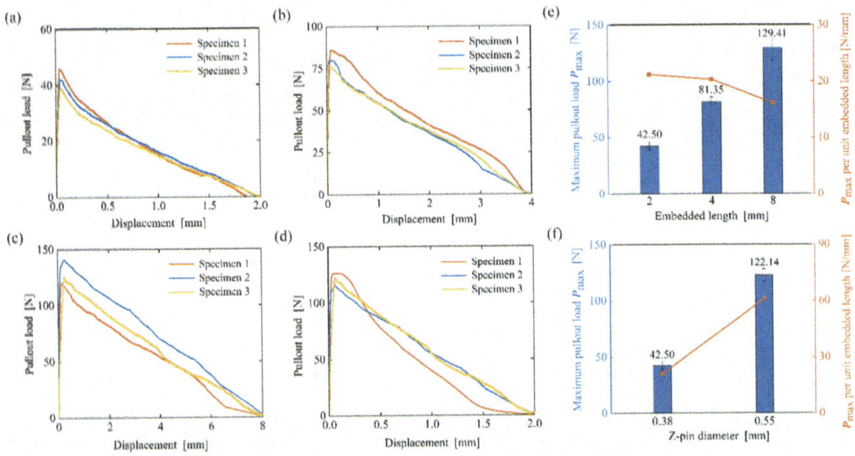

**Fig. 5.16** Load–displacement curves obtained for thin Z-pins embedded in the **a** 2 mm **b** 4 mm and **c** 8 mm through-thickness laminates; **d** thick Z-pins embedded in the 2 mm through-thickness laminates; the maximum pullout value and its per unit embedded length calculated for corresponding **e** Z-pin embedded length and **f** Z-pin diameter conditions

$$\boldsymbol{P}_i = \begin{cases} \frac{\boldsymbol{P}_{\max}}{\delta_e} \cdot \delta & (0 < \delta \le \delta_e) \\ \boldsymbol{P}_{\max} - \frac{\boldsymbol{P}_{\max}}{L_{em} - \delta_e} \cdot (\delta - \delta_e) & (\delta_e < \delta \le L_{em}) \end{cases} \tag{5.26}$$

where $L_{em}$ represents one-half through-thickness of the specimen.

It is assumed that the initial Z-pin is partially debonded to the base laminate with interfacial cracking around the perimeter ($\delta = 0$). Subsequently, during the first stage of pullout loading ($0 < \delta \le \delta_e$), elastic deformation of the Z-pin is the dominant bridging response on account of linear increase. Partially bonding adhesion plays a major role in effectively resisting the growth of delamination cracks. After reaching the limit value $\boldsymbol{P}_{max}$ ($\delta_e < \delta \le L_{em}$), the increased shear stress exceeds the interfacial shear strength, and the crack propagates seriously to other bonded locations until the processing is completed. It is known that there is no physical interfacial contact in the debonded area. Hence, the bridging traction decreases significantly with the increase of debonded region and the reduction of Z-pin embedded length in this stage.

Four essential pullout parameters are calculated to evaluate Z-pin bridging behavior via load–displacement curves. The average interfacial strength $\boldsymbol{\tau}_{ave}$, which depicts the Z-pin average pullout resistance, can be given by:

$$\boldsymbol{\tau}_{ave} = \frac{\boldsymbol{P}_{max}}{\pi d L_{em}} \tag{5.27}$$

The pullout energy $W_p$ is a function of the area under load–displacement curves and can be written as:

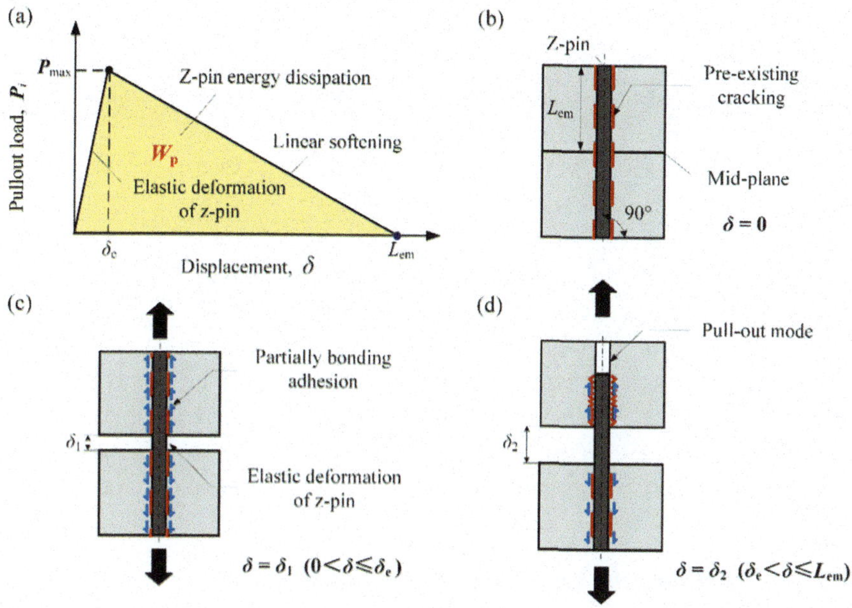


**Fig. 5.17** **a** Typical form of Z-pin bridging law, which is summarized from experimental load–displacement relationship; **b** before loading; **c** elastic loading; and **d** pullout loading

$$W_p = \int_0^{L_{em}} P_i \mathrm{d}\delta \tag{5.28}$$

Based on the determined pullout energy, the equivalent interfacial strength $\tau_{eq}$ can be defined by:

$$\tau_{eq} = \frac{2W_p}{\pi \mathrm{d} L_{em}^2} \tag{5.29}$$

Furthermore, the maximum tensile stress of the Z-pin $\sigma_{p,max}$ is determined with regard to the peak pullout force and Z-pin diameter. The definition can be expressed as:

$$\sigma_{p,max} = \frac{4P_{max}}{n\pi \mathrm{d}^2} \tag{5.30}$$

To summarize, the evaluated interfacial bond values are listed in Table 5.6.

**Table 5.6** Critical bond values of Z-pins by load–displacement data

| No | $L_{em}$ [mm] | $d$ [mm] | $P_{max}$ [N] | $\tau_{ave}$ [MPa] | $\tau_{eq}$ [MPa] | $W_p$ [N mm] | $\sigma_{p,max}$ [MPa] |
|---|---|---|---|---|---|---|---|
| 1–1 | 2 | 0.38 | 46.07 | 19.30 | 17.82 | 42.54 | 406.22 |
| 1–2 | 2 | 0.38 | 42.30 | 17.72 | 17.05 | 40.73 | 372.98 |
| 1–3 | 2 | 0.38 | 39.13 | 16.39 | 15.04 | 35.90 | 345.03 |
| 2–1 | 4 | 0.38 | 86.32 | 18.08 | 16.03 | 153.08 | 761.12 |
| 2–2 | 4 | 0.38 | 80.04 | 16.76 | 15.28 | 145.97 | 705.75 |
| 2–3 | 4 | 0.38 | 77.70 | 16.27 | 15.62 | 149.21 | 685.12 |
| 3–1 | 8 | 0.38 | 120.72 | 12.64 | 11.12 | 424.70 | 1064.44 |
| 3–2 | 8 | 0.38 | 141.42 | 14.81 | 15.07 | 575.87 | 1246.96 |
| 3–3 | 8 | 0.38 | 126.08 | 13.20 | 12.46 | 476.18 | 1111.70 |
| 4–1 | 2 | 0.55 | 126.72 | 36.67 | 25.35 | 87.59 | 533.37 |
| 4–2 | 2 | 0.55 | 116.09 | 33.59 | 32.80 | 113.34 | 488.63 |
| 4–3 | 2 | 0.55 | 123.61 | 35.77 | 33.24 | 114.88 | 520.28 |

## 5.3.2 Fractography and Failure Mechanism

The failure photomicrographs of Z-pins after pullout testing are illustrated in Fig. 5.18. Notably, resin-rich warping is pronounced nearest the Z-pin. All Z-pins are pulled out completely from the base laminates, regardless of different Z-pinning conditions. Evidently, this failure mode confirms the inferior interfacial adhesion produced by the cure. This suggests that the pre-existing interfacial cracking should be considered as a significant factor further exploring its effect on the bridging traction capacity.

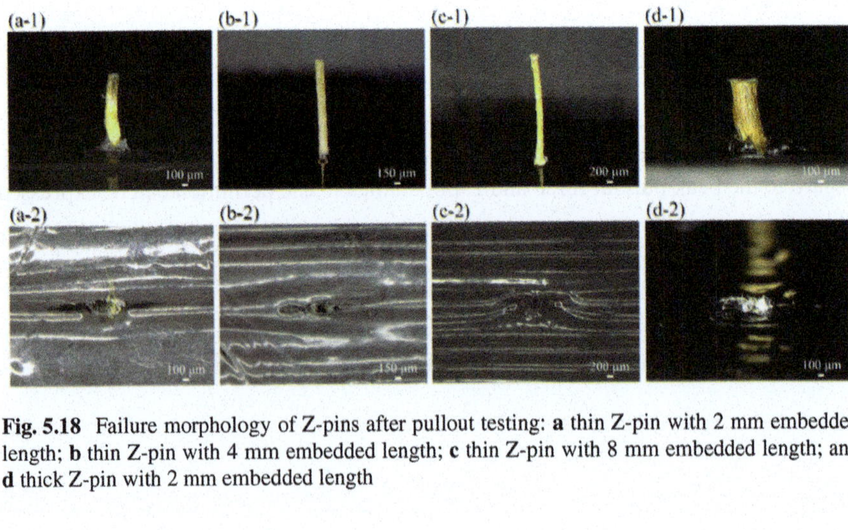


**Fig. 5.18** Failure morphology of Z-pins after pullout testing: **a** thin Z-pin with 2 mm embedded length; **b** thin Z-pin with 4 mm embedded length; **c** thin Z-pin with 8 mm embedded length; and **d** thick Z-pin with 2 mm embedded length

### 5.3.3 Cure Reaction Properties

A phenomenological model is typically proposed for resin kinetics to capture its conversion-dependent reaction mechanics throughout the cure [22]. The curing rate $\frac{d\alpha}{dt}$ is related to the curing temperature $T$ and degree of cure (DoC) $\alpha$, which is shown as:

$$\frac{d\alpha}{dt} = f(T, \alpha) \tag{5.31}$$

Based on Johnston et al. [23], the nonlinear relation $f(T,\alpha)$ between the curing temperature and DoC for the epoxy is described by:

$$f(T, \alpha) = \frac{k(T)\alpha^m(1-\alpha)^n}{1+\exp[C(\alpha-\alpha_c)]}, \ \alpha_c = \alpha_{c0} + \alpha_{c1}T \tag{5.32}$$

where $k(T)$ is the coefficient of reaction rate and governed by:

$$k(T) = A\exp\left(\frac{-\Delta E}{RT}\right) \tag{5.33}$$

where $m$ and $n$ are the reaction exponents, $C$ represents the diffusion constant, $\alpha_{c0}$ and $\alpha_{c1}$ are the fitting constants of DoC depending on temperature evolution, $A$ is the frequency factor, $\Delta E$ is the activation energy, and $R$ is the gas constant.

With respect to the transform from an initial viscous liquid state to an intermediate rubbery solid state, and to a final glassy solid state in the cure, the isotropic resin matrix is modeled as a CHILE material [24]. Accordingly, the elastic constants $E_r$ can be written as:

$$E_r = \begin{cases} E_r^0 & T^* < T_{C1}^* \\ E_r^0 + \frac{T^*-T_{C1}^*}{T_{C2}^*-T_{C1}^*}\left(E_r^\infty - E_r^0\right) & T_{C1}^* \le T^* < T_{C2}^* \\ E_r^\infty & T_{C2}^* \le T^* \end{cases} \tag{5.34}$$

where $E_r^0$ and $E_r^\infty$ denote the initial liquid and final solid moduli, respectively, $T_{C1}^*$ and $T_{C2}^*$ are the lower and upper fitting constants with increase in temperature, $T^*$ represents the temperature difference between glass transition and instantaneous value, i.e., $T^* = T_g - T$. The mathematical formulas of these critical parameters can be totally derived from Johnston's work in great detail [9] and respectively written by:

$$T_g = T_g^0 + a_{Tg}\cdot\alpha, \ T_{C1}^* = T_{C1a}^* + T_{C1b}^*\cdot T \tag{5.35}$$

where $T_g^0$ is the glass transition temperature at $\alpha = 0$, $a_{Tg}$ is the linear coefficient of glass transition temperature with $\alpha = 0$, $T_{C1a}^*$ and $T_{C1b}^*$ are the two critical temperature points.

Since resin matrix is a mixture of uncured (liquid and rubbery) and fully cured (hard glass) states in the cure, the coefficient of thermal expansion (CTE) is defined via rule of mixture (RoM) method:

$$\beta_r(\alpha) = (1-\alpha)\beta_r^0 + \alpha\beta_r^\infty \tag{5.36}$$

where ${\beta_r}^0$ and $\beta_r^\infty$ represent the CTEs of initially uncured and completely cured states, respectively.

Also, chemical shrinkage considerably occurs in the thermosetting resin as a result of increase in density and corresponding reduction of volume for cross-linking. The coefficient of chemical shrinkage (CCS) is introduced to describe the shrinkage variation [25].

The homogenized properties of the laminated composite can be characterized by fiber and resin matrix components in terms of self-consistent micromechanics approach [26]. The mechanical moduli $E$ together with Poisson's ratios $v$ following the longitudinal and transversal directions are expressed by:

$$\begin{cases} E_{11} = E_{f1}V_f + E_r(1-V_f) + \left[\dfrac{4(v_r - v_{f12}^2)K_f K_r G_r(1-V_f)V_f}{(K_f+G_r)K_r + (K_f-K_r)G_r V_f}\right] \\ E_{22} = E_{33} = \dfrac{1}{\frac{1}{4K} + \frac{1}{4G_{23}} + \frac{v_{12}^2}{E_{11}}} \\ G_{12} = G_{13} = G_r\left[\dfrac{(G_{f12}+G_r) + (G_{f12}-G_r)V_f}{(G_{f12}+G_r) - (G_{f12}-G_r)V_f}\right] \\ G_{23} = G_r\left[\dfrac{K_r(G_{f23}+G_r) + 2G_r G_{f23} + K_r(G_{f23}-G_r)V_f}{K_r(G_{f23}+G_r) + 2G_r G_{f23} - (K_r+2G_r)(G_{f23}-G_r)V_f}\right] \\ v_{12} = v_{13} = v_{f12}V_f + v_r(1-V_f) + \left[\dfrac{(v_r - v_{f12})(K_r-K_f)G_r(1-V_f)V_f}{(K_f+G_r)K_r + (K_f-K_r)G_r V_f}\right] \\ v_{23} = \dfrac{2E_{11}K - E_{11}E_{22} - 4v_{12}^2 E_{22}K}{2E_{11}K} \end{cases} \tag{5.37}$$

where the subscripts 1, 2 and 3 denote three principal ply directions, f and r are on behalf of the fiber and resin components, $V_f$ is the volume content of fibers within the composite, and $K$ is the bulk modulus governed by:

$$\begin{cases} K_r = \dfrac{E_r}{2(1-v_r-2v_r^2)} \\ K_f = \dfrac{E_f}{3(1-2v_f)} \\ K = \dfrac{K_r(K_f+G_r) + G_r V_f(K_f-K_r)}{(K_f+G_r) - V_f(K_f-K_r)} \end{cases} \tag{5.38}$$

The integrated CTEs of carbon/epoxy composite determined by RoM method are given by:

$$\begin{cases} \beta_1 = \frac{\beta_{f1} E_{f1} V_f + \beta_r E_r (1 - V_f)}{E_{f1} V_f + E_r (1 - V_f)} \\ \beta_2 = \beta_3 = (\beta_{f2} + \nu_{f12} \beta_{f1}) V_f + (\beta_r + \nu_r \beta_r)(1 - V_f) \\ \quad -[\nu_{f12} V_f + \nu_r (1 - V_f)]\beta_1 \end{cases} \tag{5.39}$$

Similarly, the homogenization of CCSs along the three principal ply directions can be written as:

$$\begin{cases} \gamma_1 = \dfrac{E_r (1 - V_f)\gamma_r}{E_{f1} V_f + E_r (1 - V_f)} \\ \gamma_2 = \gamma_3 = (1 + \nu_r)(1 - V_f)\gamma_r - [\nu_{f12} V_f + \nu_r (1 - V_f)]\gamma_1 \end{cases} \tag{5.40}$$

As a consequence of differences in thermal expansion and curing shrinkage between fiber and matrix within a ply, a micromechanical residual stress $\boldsymbol{\sigma}$ is developed throughout the cure. A stress calculation is determined to evaluate this variable as follows:

$$\begin{cases} \boldsymbol{\sigma} = \boldsymbol{C}\varepsilon_{mech} \\ \varepsilon_{mech} = \varepsilon_{total} - \varepsilon_{th} - \varepsilon_{sh} \end{cases} \tag{5.41}$$

where $\boldsymbol{C}$ is the stiffness tensor of chosen materials associated with mechanical moduli, $\varepsilon_{total}$ is the total strain which can be decomposed into the mechanical part $\varepsilon_{mech}$, thermal part $\varepsilon_{th}$, and shrinkage part $\varepsilon_{sh}$. The $\varepsilon_{th}$ and $\varepsilon_{sh}$ related to the curing temperature and DoC are respectively given by:

$$\begin{cases} \varepsilon_{th} = \beta \cdot T \\ \varepsilon_{sh} = \gamma \cdot \alpha \end{cases} \tag{5.42}$$

### 5.3.4 Finite Element Implementation

An RVE-based finite element method is proposed to gain a detailed understanding of CE on the bridging mechanism of Z-pins within carbon/epoxy laminates. The geometry of the RVE, based on the characterization, is designed as 3 mm × 3 mm × 2 mm for the thin Z-pin (0.38 mm) and 6 mm × 6 mm × 2 mm for the thick Z-pin (0.55 mm), respectively. The prepreg ply is arranged in $[0]_{32}$ configurations. The base laminates and resin-rich zone are discretized via 8-node brick elements (C3D8). The Z-pin and interface regions are modeled employing cohesive elements (COH3D8). For verification, the stacking consequences, material types and curing cycles applied to the RVE are consistent with the experimental configurations. According to the previous experimental data, the curing properties and damage parameters of the RVE are listed in Tables 5.7 and 5.8, respectively.

An integrated framework between the cure reaction and subsequent pullout testing is provided in Fig. 5.19, wherein a three-step coupled simulation is conducted: (1) a thermal-chemical analysis is carried out for the quantification of residual stresses during the cure. The cure-dependent constitutive properties for each component are characterized using Abaqus™ external subroutines. The obtained temperature, DoC

**Table 5.7** Thermochemical properties of T300/8552 reinforced with Polyimide Z-pin

| Components | Categories | | |
|---|---|---|---|
| Fiber | Elastic modulus | $E_{f1} = 230{,}000$ MPa | $E_{f2} = E_{f3} = 13{,}800$ MPa |
| | Shear modulus | $G_{f12} = G_{f23} = 8970$ MPa | $G_{f23} = 4380$ MPa |
| | Poisson's ratio | $\nu_{f12} = \nu_{f13} = 0.2$ | $\nu_{f23} = 0.25$ |
| | CTE | $\beta_{f1} = -0.7 \times 10^{-6}\ \mathrm{K}^{-1}$ | $\beta_{f2} = \beta_{f3} = 12 \times 10^{-6}\ \mathrm{K}^{-1}$ |
| Resin | Elastic modulus | $E_r^0 = 4.08$ MPa | $E_r^\infty = 4080$ MPa |
| | Poisson's ratio | $\nu_r = 0.38$ | |
| | Kinetics | $A = 70{,}000\ \mathrm{s}^{-1}$ | $m = 0.5$ |
| | | $n = 1.5$ | $C = 30$ |
| | | $\alpha_{c0} = -1.515$ | $\alpha_{c1} = 5.171 \times 10^{-3}\ \mathrm{K}^{-1}$ |
| | | $\Delta E = 65{,}000$ J/mol | $R = 8.3145\ \mathrm{J/(mol\ K)}^{-1}$ |
| | | $T_g^0 = 268$ K | $T_{C1}^* = -45.7$ K |
| | | $T_{C2}^* = -12$ K | $a_{Tg} = 220$ K |
| | CTE | $\beta_r^0 = 81.5 \times 10^{-6}\ \mathrm{K}^{-1}$ | $\beta_r^\infty = 32.6 \times 10^{-6}\ \mathrm{K}^{-1}$ |
| | CCS | $\gamma_r = -0.055$ | |
| Z-pin | Elastic modulus | $E_{11} = 100{,}000$ MPa | $E_{22} = E_{33} = 30{,}000$ MPa |
| | Shear modulus | $G_{12} = G_{13} = 4600$ MPa | $G_{23} = 3900$ MPa |
| | Poisson's ratio | $\nu_{12} = \nu_{13} = 0.35$ | $\nu_{23} = 0.4$ |

**Table 5.8** Critical damage parameters for Z-pin, resin matrix, and interface

| Components | Categories | | | | | |
|---|---|---|---|---|---|---|
| | Thin | | | Thick | | |
| Z-pin | $\sigma_{p,n}^0$ | $\sigma_{p,s}^0$ | $\sigma_{p,t}^0$ | $\sigma_{p,n}^0$ | $\sigma_{p,s}^0$ | $\sigma_{p,t}^0$ |
| | 500 MPa | 500 MPa | 500 MPa | 600 MPa | 600 MPa | 600 MPa |
| Resin | $\kappa$ | $\psi$ | $\bar{\varepsilon}_{0-}^{pl}$ | $\bar{\varepsilon}_{0+}^{pl}$ | $G_D$ | $k$ |
| | 97.6 MPa | 24° | 0.25 | 0.025 | 0.001 N/mm | 0.8 |
| Interface | $K_{nn} = K_{ss} = K_{tt}$ | $\sigma_n^0$ | $\sigma_s^0 = \sigma_t^0$ | $G_{IC}$ | $G_{IIC} = G_{IIIC}$ | $\eta$ |
| | $10^5\ \mathrm{N/mm}^3$ | 50 MPa | 25 MPa | 0.01 N/mm | 1 N/mm | 1 |

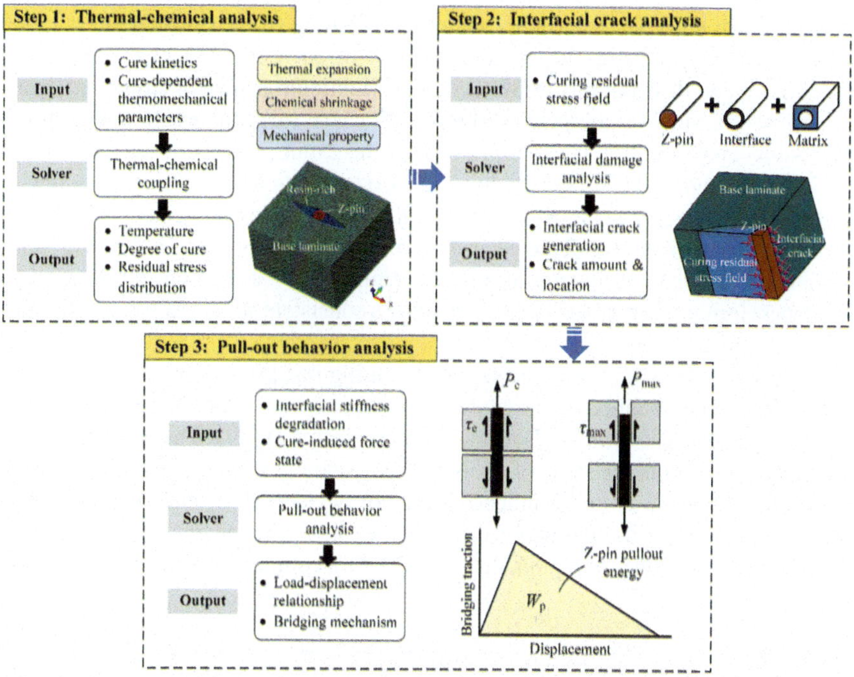


**Fig. 5.19** Integrated framework linking the cure process and subsequent pullout behavior of a single Z-pinned carbon/epoxy laminate

and residual stress distributions are regarded as an initial boundary in the following steps; (2) in terms of interfacial damage computation, the specific location and amount of debonding produced by high-stress concentration are explicitly evaluated; (3) a subsequent pullout-behavior analysis is adopted to identify the traction properties of Z-pins, with predicted debonding into account. A combination of interfacial stiffness degradation and cure-induced force states is applied to the RVE to represent the curing equivalent characterization. The simulated load–displacement relations are validated against the experimental data.

### 5.3.5 *Cure-Induced Residual Stress Computation*

Via thermal-chemical simulation, the predictions for the thin Z-pin are shown in Fig. 5.20. The evolutions of temperature and DoC under the recommended curing cycles are plotted in Fig. 5.20a. The DoC reaches 0.93 before the cooling-down stage and remains the constant value until the process is complete. As illustrated in Fig. 5.20b, the evolutions of curing residual stresses for each component (including base laminate, resin-rich, Z-pin and interface) are respectively noted. Clearly, the cooling-down stage is the primary factor contributing to the residual stress development. For comparison, the maximum stress of each component is collected in Fig. 5.20c. The distributions of Mises stress, Maximum principal stress and Transverse stress in the RVE are displayed in Fig. 5.20d–f, respectively. Remarkably, the high-stress concentration is generated around the Z-pin. The resin-rich neighboring the Z-pin exhibits a tensile state, while the base laminate nearby the resin-rich withstands a compressive state. More specifically, the stresses generated in the Z-pin and interface regions are depicted in Fig. 5.20g–i, respectively. Due to the unidirectional stacking consequence within the laminate, the stress concentration of the Z-pin develops along a straight path, providing a consistent stress track to the interface pattern.

In the case of thick Z-pins, the simulated curing results are explored, as depicted in Fig. 5.21. The obtained temperature and DoC curves are consistent with the thin-diameter results. The ultimate DoC maintains the value of 0.93. The residual stresses are greatly created in the stage of cooling-down, and the interface displays the high-stress concentration. The cure-induced force state shows similar characteristics at the micro-scale. Obviously, the major difference is the improvement of residual stress developed in the interface region, up to + 130.40% compared to the thin Z-pin case.

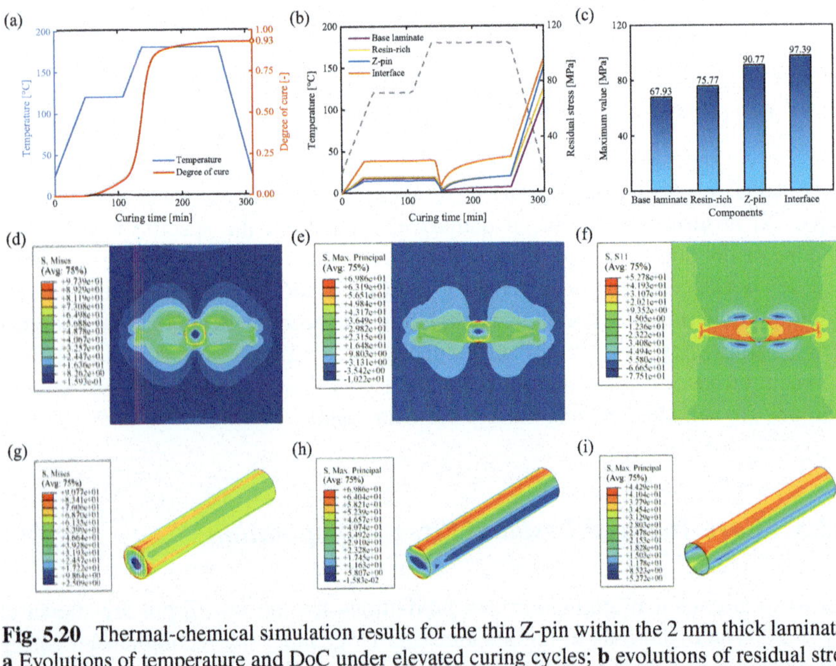


**Fig. 5.20** Thermal-chemical simulation results for the thin Z-pin within the 2 mm thick laminates. **a** Evolutions of temperature and DoC under elevated curing cycles; **b** evolutions of residual stress and **c** maximum stress value for each component in the RVE; distributions of **d** Mises stress, **e** maximum principal stress and **f** transverse stress in the RVE; distributions of **g** Mises stress and **h** Maximum principal stress in the Z-pin; **i** distribution of Maximum principal stress in the interface

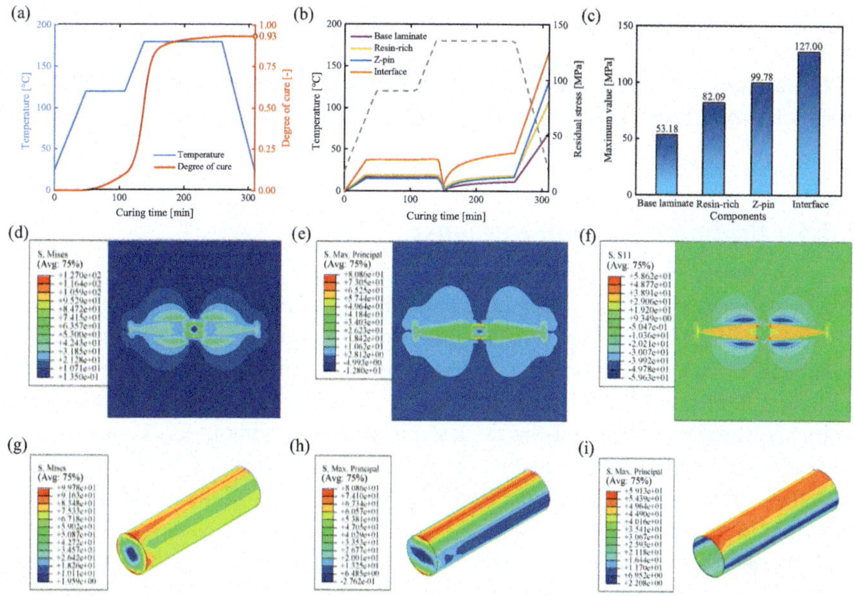


**Fig. 5.21** Thermal-chemical simulation results for the thick Z-pin within the 2 mm thick laminates. **a** Evolutions of temperature and DoC under elevated curing cycles; **b** evolutions of residual stress and **c** maximum stress value for each component in the RVE; distributions of **d** Mises stress, **e** maximum principal stress and **f** transverse stress in the RVE; distributions of **g** Mises stress and **h** maximum principal stress in the Z-pin; **i** distribution of Maximum principal stress in the interface

The failure responses of the interface caused by high residual stresses are depicted in Fig. 5.22. Significantly, the local stress concentration dramatically exceeds the interfacial strength, indicative of terrible interfacial debonding prior to loading. The factor of debonding area ratio $\xi_d$ is implemented here for the quantification of cracking, which can be defined as follows:

$$\xi_d = \frac{A_d}{A_0} \times 100\% \tag{5.43}$$

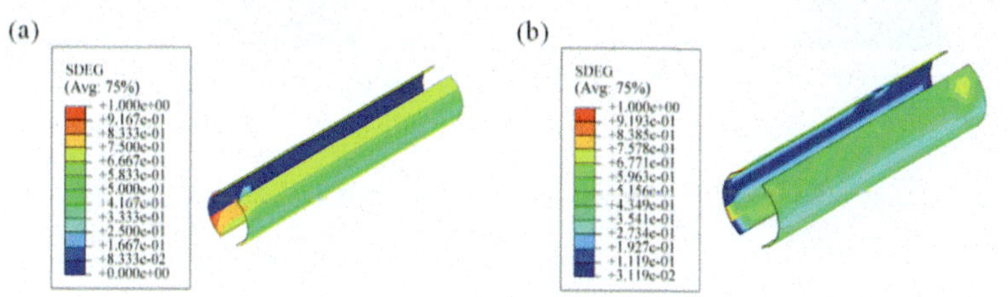


**Fig. 5.22** Interfacial crack simulation results for **a** thin and **b** thick Z-pins within the 2 mm thick laminates

where $A_d$ is the area of interfacial cracking, $A_0$ is the whole area of the interface including bonding and debonding.

Consequently, the ultimate $\xi_d$ is estimated as 20% for the thin Z-pin and 15% for the thick Z-pin, respectively.

### *5.3.6 Bridging Traction Response*

The bridging performance of Z-pins accounting for the debonding crack is explicitly estimated. The simulated load–displacement relations with and without CE are compared in Fig. 5.23. For the thin Z-pin, the maximum pullout force with and without CE is predicted to be 41.26 N and 51.33 N, respectively. Similarly, the estimated maximum value of the thick Z-pin with and without CE is estimated as 126.78 N and 146.68 N, respectively. The results considering the CE coincide closely with the experiments, denoting the accuracy and reliability of the modeling strategy proposed in this study.

A mathematical relation between the maximum pullout force and partial debonding of the Z-pin can be given. It is known that debonding area provides no physical interfacial adhesion, thus significantly reducing Z-pin bridging tractions. Accordingly, assuming that the traction is linearly associated with the bonding area, the maximum pullout load can be governed by:

$$\boldsymbol{P}_{\max} = \boldsymbol{\tau}_{\mathrm{b}} \pi \mathrm{d} L_{\mathrm{em}} \cdot (1 - \xi_{\mathrm{d}}) \tag{5.44}$$

where $\tau_b$ is the fully bonding strength of the interface.

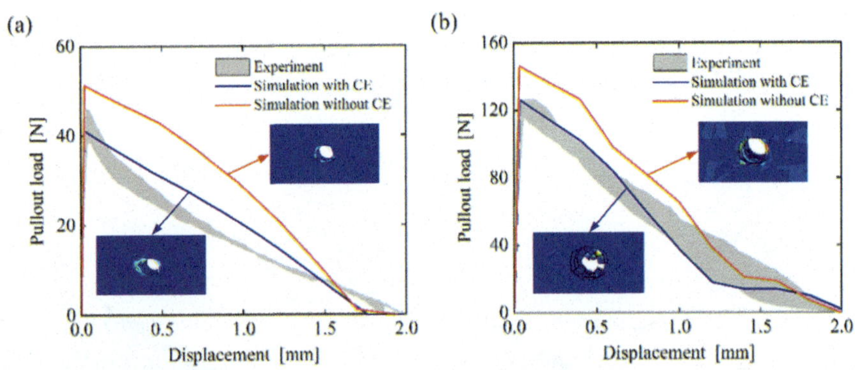


**Fig. 5.23** Pullout behavior simulation results for **a** thin and **b** thick Z-pins

### 5.3.7 *Influence of Holding Temperature on the Z-pin Bridging*

A parametric investigation of the critical curing parameters is carried out for the thin Z-pin reinforcement. The contributing factors, including holding temperature, holding time and heating rate of the recommended curing cycles, are changed, respectively, as drawn in Fig. 5.24. In detail, three redesigned temperature cycles with the original are compared in Table 5.9.

Compared with Case 1, the second holding temperature in Case 2 is decreased from 180 to 160 °C. The simulated curing defects and bridging response of Z-pin are drawn in Fig. 5.25, with a comparison of Case 1. The ultimate DoC is decreased from 0.93 to 0.83, indicating the inhibition of heat release and efficiency of the cure reaction. The reduction of holding temperature produces a decreased modulus of resin, thus resulting in a low curing mismatch within the cure. Hence, the inhibition of the curing mismatch relieves residual stress development. The stress is mainly concentrated around the perimeter of the Z-pin with a great decrease. Remarkably, the maximum value of interfacial stress in Case 2 is reduced to 95.42 MPa. The simulated load–displacement curve in Case 2 also exhibits a bi-linear trend, and the

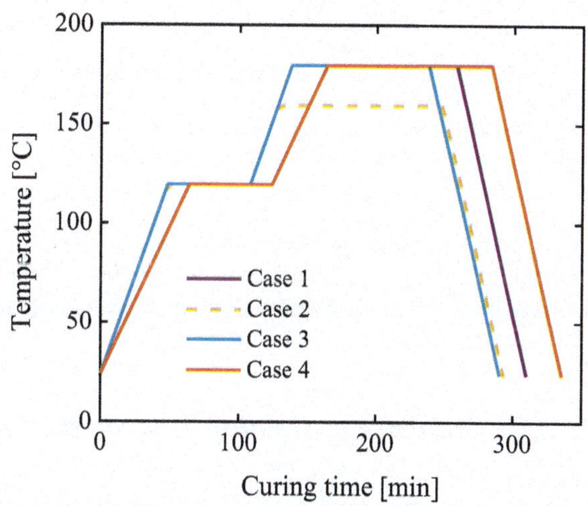


**Fig. 5.24** Three redesigned curing cycles compared with the original one for the parametric investigation

**Table 5.9** Comparison of curing temperature conditions

| | Holding temperature (°C) | Holding time (min) | Heating rate (°C /min) |
|---|---|---|---|
| Case 1 | 180 | 120 | 2 |
| Case 2 | 160 | 120 | 2 |
| Case 3 | 180 | 90 | 2 |
| Case 4 | 180 | 120 | 1.5 |

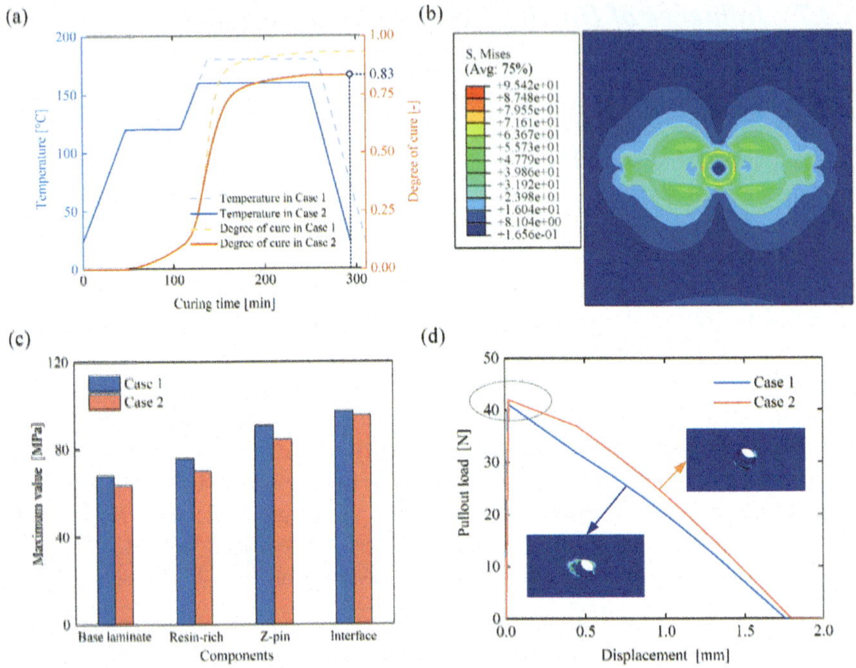


**Fig. 5.25** Simulation results in Case 2 compared with Case 1. **a** Evolutions of temperature and DoC under redesigned elevated curing cycles; **b** distribution of curing residual stresses in the RVE; **c** maximum stress value obtained for each component after the cure; **d** variations of pullout load–displacement relationship under two curing conditions

maximum pullout value is increased to 42.41 N due to reduced stress and interfacial cracking development. The area under the curve is improved obviously, which means an amelioration of the Z-pin pullout energy under external loading.

### 5.3.8 *Influence of Holding Time on the Z-pin Bridging*

With the comparison of Case 1, the holding time is decreased from 120 to 90 min. The predicted numerical results, involving curing predictions and Z-pin pullout behavior, are explicitly explored, as denoted in Fig. 5.26. Notably, the final DoC is limited to 0.92 with a paltry decrease in Case 3. In the absence of holding time, the DoC stops improving and maintains a slightly lower value until the processing is completed. The overall curing residual stresses are slightly limited. This is owing that the shortage of holding time triggers an incomplete curing reaction and non-uniform distribution of residual stresses around the Z-pin. The maximum value of interfacial stress is slightly improved to 98.05 MPa, and the highest residual stress exists indeed in the vicinity of Z-pin throughout the cure. Significantly, the maximum pullout load is

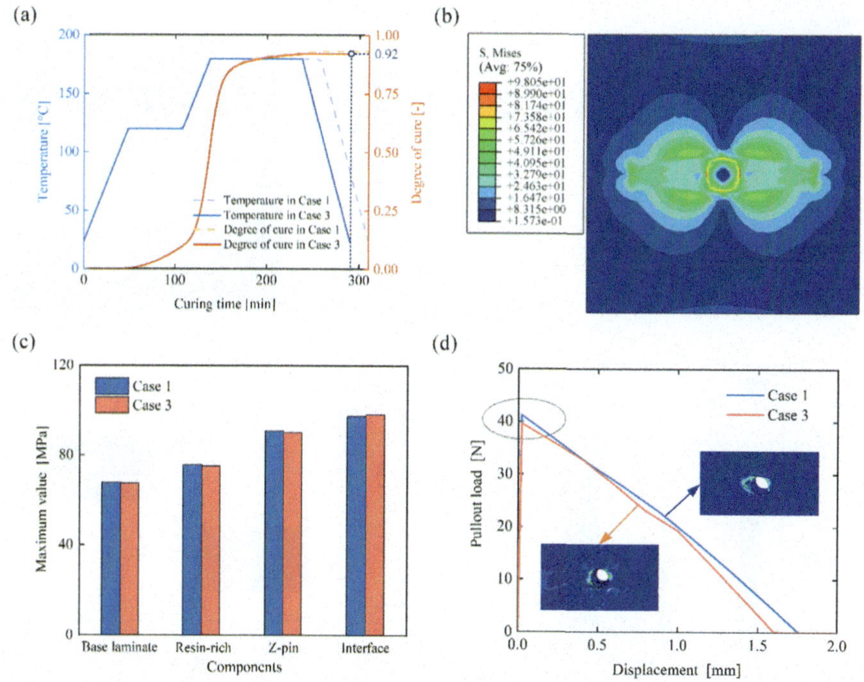


**Fig. 5.26** Simulation results in Case 3 compared with Case 1. **a** Evolutions of temperature and DoC under redesigned elevated curing cycles; **b** distribution of curing residual stresses in the RVE; **c** maximum stress value obtained for each component after the cure; **d** variations of pullout load–displacement relationship under two curing conditions

diminished to 39.25 N and the displacement is lessened, producing a sharp drop in Z-pin pullout energy. This reveals that the relatively short holding time results in the suppression of interfacial residual stress development, thus seriously impairing the bridging efficiency of Z-pins.

### 5.3.9 *Influence of Heating Rate on the Z-pin Bridging*

Compared with Case 1, the heating rate in Case 4 is decreased from 2 to 1.5 °C/min. The prediction of CE on the Z-pin bridging behavior is denoted in Fig. 5.27. The evolutions of DoC are changed dramatically, and the ultimate value reaches 0.94. Appreciably, the increase in the DoC of Case 4 slows down as the heating rate decreases. The slower heating rate leads to reduced temperature difference per cure reaction increment, thus contributing to the sufficiency of cure reaction during the processing. Due to the relatively long curing time, the chemical shrinkage of the resin becomes more uniform. This phenomenon alleviates the partial stress concentration between different phases and effectively reduces the residual stresses. Therefore, the

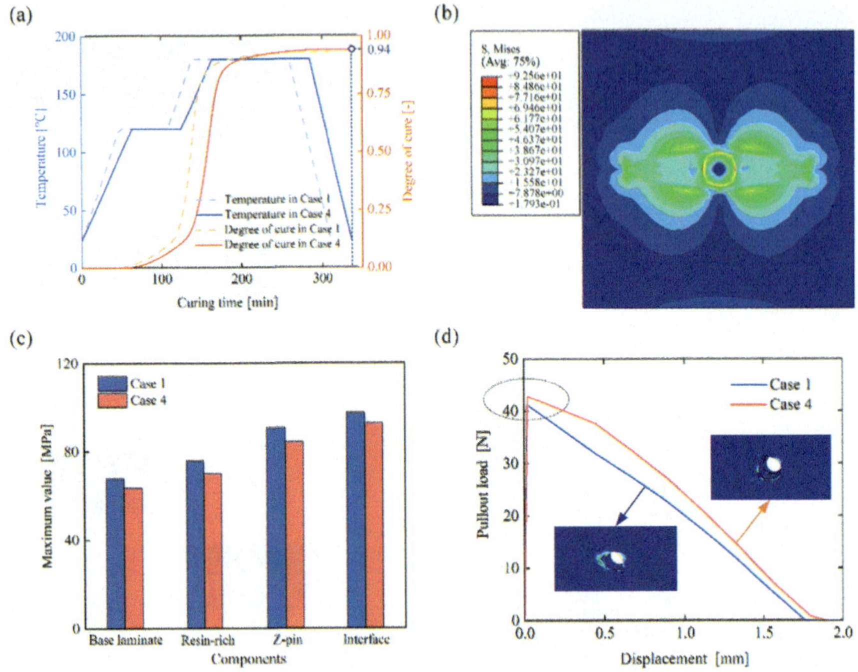


**Fig. 5.27** Simulation results in Case 4 compared with Case 1. **a** Evolutions of temperature and DoC under redesigned elevated curing cycles; **b** distribution of curing residual stresses in the RVE; **c** maximum stress value obtained for each component after the cure; **d** variations of pullout load–displacement relationship under two curing conditions

maximum stress value is significantly cut down for each component. Outstandingly, the highest stress remains nearby the Z-pin. The final interfacial stress is decreased to 92.56 MPa. According to this, the development of interfacial cracking is also limited, and the physical interfacial contact of Z-pin is improved considerably. The maximum pullout force is pronouncedly increased to 43.45 N. The Z-pin pullout energy is greater than that of Case 1, being sufficiently effective at resisting the delamination crack.

## 5.4 Conclusions

In this chapter, a novel micromechanical model is presented to disclose the influence of curing process on the mechanical properties of Z-pinned CFRP composites. The fiber waviness and resin-rich pockets triggered by Z-pins are considered during the RVE model generation. The variations of material properties during the cure are characterized using CHILE model. The effect of Z-pinning parameters on the residual stress development is investigated. The properties with and without CRS are obtained at different Z-pin conditions. The predicted results show good agreement with the experimental data.

Cure-induced residual stresses of the Z-pinned CFRP composites cannot be ignored due to maximum stresses formed in the interface. The increased Z-pin density promotes continuous resin-rich channels and is responsible for the weakening mechanical properties in the vicinity of the Z-pins after the cure, due to increased interfacial stress concentration. Both interfacial stress concentration and continuous resin-rich channels that adversely affect the mechanical properties are responsible for the reduction of strength, elastic modulus, and transformation of failure mode in the cured Z-pinned CFRP composites. The obtained strength and failure strain are slightly increased at low Z-pin density but strongly decreased at intermediate Z-pin density and high Z-pin density, comparing the simulated results with and without CRS. In addition, the inherent curing mismatch is responsible for the generation of interfacial cracking around the Z-pins. The presence of cracking leads to a limitation of interfacial bonding, thereby adversely affecting the bridging adhesion of the Z-pins. The influence of curing parameters on the traction ability of Z-pins is further explored, providing significant findings to promote high-quality control of Z-pinned CFRP composites.

To summarize, the occurrence of microstructural defects caused by the cure, including residual stress concentration and interfacial cracking, is an inherent problem for the Z-pin enhancement. An appropriate design of Z-pinning parameters and curing parameters is quite necessary and should be identified to ensure high mechanical performance of Z-pin enhancement.

## References


1. Yasaee M, Bigg L, Mohamed G, Hallett SR. Influence of Z-pin embedded length on the interlaminar traction response of multi-directional composite laminates. Material Design. 2017;115:26–36.
2. Yan W, Liu HY, Mai YW. Numerical study on the mode I delamination toughness of Z-pinned CFRP composites. Composites Science and Technology. 2003;63:1481–93.
3. Yan W, Liu HY, Mai YW. Mode II delamination toughness of Z-pinned CFRP composites. Composites Science and Technology. 2004;64:1937–45.
4. Grassi M, Zhang X. Finite element analyses of mode I interlaminar delamination in z-fibre reinforced composite laminates. Composites Science and Technology. 2003;63:1815–32.
5. Fu C, Wang Xi. Micro-mechanical analysis of matrix crack-induced delamination in cross-ply laminates in tension. Composite Structures. 2020;243:112202.
6. Zhang B, Allegri G, Yasaee M, Hallett S. Micro-mechanical finite element analysis of Z-pins under mixed-mode loading. Composites Part A: Applied Science and Manufacturing. 2015;78:424–35.
7. Hoffmann J, Scharr G. Mechanical properties of composite laminates reinforced with rectangular Z-pins in monotonic and cyclic tension. Composites Part A: Applied Science and Manufacturing. 2018;109:163–70.
8. Isa MD, Feih S, Mouritz AP. Compression fatigue properties of Z-pinned quasiisotropic carbon/epoxy laminate with barely visible impact damage. Composite Structures. 2011;93(9):2269–76.
9. Johnston A. An Integrated Model of the Development of Process-Induced Deformation in Autoclave Processing of Composite Structures. University of British Columbia. 1998.

10. Maiarù M, D'Mello RJ, Waas AM. Characterization of intralaminar strengths of virtually cured polymer matrix composite. Composites Part B: Engineering. 2018;149:285–95.
11. Li D, Li X, Dai J, Xi S. A Comparison of Curing Process-Induced Residual Stresses and Cure Shrinkage in Micro-Scale Composite Structures with Different Constitutive Laws. Applied Composite Materials. 2018;25(1):67–84.
12. Whitney JM. Elastic Moduli of Unidirectional Composites with Anisotropic Filaments. Journal of Composite Materials. 1967;1(2):188–93.
13. Vignoli L, Savi M, Pacheco P, Kalamkarov A. Comparative analysis of micromechanical models for the elastic composite laminae. Composites Part B: Engineering. 2019;174:106961.
14. Abou msallem Y, Jacquemin F, Boyard N, Poitou A, Delaunay D, Chatel S. Material characterization and residual stresses simulation during the manufacturing process of epoxy matrix composite. Composites Part A Applied Science and Manufacturing. 2010;41:108–15.
15. Chang P, Mouritz AP, Cox BN. Properties and failure mechanisms of Z-pinned CFRP composites in monotonic and cyclic tension. Composites Part A Applied Science and Manufacturing. 2006;37(10): 1501–13.
16. Tran TD, Kelly D, Prusty BG, Gosse JH, Christensen S. Micromechanical modelling for onset of distortional matrix damage of fiber reinforced composite materials. Composite Structures. 2012;94(2):745–57.
17. Hui X, Xu Y, Zhang W. An integrated modeling of the curing process and transverse tensile damage of unidirectional CFRP composites. Composite Structures. 2021;263: 113681.
18. Camanho PP, Davila CG, de Moura MF. Numerical simulation of mixed-mode progressive delamination in composite materials. Journal of Composite Materials 2003;37(16): 1415–38.
19. Gu J, Chen P. Some modifications of Hashin's failure criteria for unidirectional composite materials. Composite Structures. 2017;182:143–52.
20. Suquet P. Elements of homogenization for inelastic solid mechanics. Homogenization for Composite Materials. 1987;272:193–278.
21. Yasaee M, Lander JK, Allegri G, Hallett S. Experimental characterisation of mixed mode traction-displacement relationships for a single carbon composite Z-pin. Composites Science and Technology. 2014;94:123–131.
22. Ersoy N, Tugutlu MJPE, Science. Cure kinetics modeling and cure shrinkage behavior of a thermosetting composite. Material Science. 2010;50:84–92.
23. Johnston A, Vaziri R, Poursartip A. A plane strain model for process-induced deformation of laminated composite structures. Journal of Composite Materials. 2001;35:1435–69.
24. Hui X, Xu Y, Wang J, Zhang W. Microscale viscoplastic analysis of unidirectional CFRP composites under the influence of curing process. Composite Structures. 2021;266:113786.
25. Kravchenko OG, Kravchenko SG, Pipes RB. Chemical and thermal shrinkage in thermosetting prepreg. Composites Part A: Applied Science and Manufacturing. 2016;80:72–81.
26. Zhang G, Wang J, Ni A, Li S. Process-induced residual stress of variable-stiffness composite laminates during cure. Composite Structures. 2018;204:12–21.

The manufacturer's authorised representative in the EU is Springer Nature Customer Service Centre GmbH, Europaplatz 3, 69115 Heidelberg, Germany. If you have any concerns regarding our products, please contact ProductSafety@springernature.com

Printed and bound by CPI Group (UK) Ltd, Croydon, CR0 4YY
30/09/2026
02225265-0008